Global Transformations in Media and Communication Research - A Palgrave and IAMCR Series

Series Editors

Marjan de Bruin
HARP, Mona Campus
The University of the West Indies HARP, Mona Campus
Mona, Jamaica

Claudia Padovani
SPGI
University of Padova
Padova, Padova, Italy

The International Association for Media and Communications Research (IAMCR) has been, for over 50 years, a focal point and unique platform for academic debate and discussion on a variety of topics and issues generated by its many thematic Sections and Working groups (see http://iamcr.org/) This new series specifically links to the intellectual capital of the IAMCR and offers more systematic and comprehensive opportunities for the publication of key research and debates. It will provide a forum for collective knowledge production and exchange through trans-disciplinary contributions. In the current phase of globalizing processes and increasing interactions, the series will provide a space to rethink those very categories of space and place, time and geography through which communication studies has evolved, thus contributing to identifying and refining concepts, theories and methods with which to explore the diverse realities of communication in a changing world. Its central aim is to provide a platform for knowledge exchange from different geo-cultural contexts. Books in the series will contribute diverse and plural perspectives on communication developments including from outside the Anglo-speaking world which is much needed in today's globalized world in order to make sense of the complexities and intercultural challenges communication studies are facing.

More information about this series at
http://www.palgrave.com/gp/series/15018

Joana Díaz-Pont • Pieter Maeseele
Annika Egan Sjölander
Maitreyee Mishra • Kerrie Foxwell-Norton
Editors

The Local and the Digital in Environmental Communication

palgrave
macmillan

Editors
Joana Díaz-Pont
Department of Political Science and Public Law
Autonomous University of Barcelona
Bellaterra, Spain

Pieter Maeseele
Department of Communication Studies
University of Antwerp
Antwerp, Belgium

Annika Egan Sjölander
Department of Culture and Media Studies
Umeå University
Umeå, Sweden

Maitreyee Mishra
Manipal Institute of Communication
Manipal Academy of Higher Education (Manipal University)
Manipal, India

Kerrie Foxwell-Norton
School of Humanities, Languages and Social Science
Griffith University
Gold Coast, QLD, Australia

Global Transformations in Media and Communication Research - A Palgrave and IAMCR Series
ISBN 978-3-030-37329-0 ISBN 978-3-030-37330-6 (eBook)
https://doi.org/10.1007/978-3-030-37330-6

Cover illustration: TimeStopper / gettyimages
Cover design: eStudioCalamar

This Palgrave Macmillan imprint is published by the registered company Springer Nature Switzerland AG.
The registered company address is: Gewerbestrasse 11, 6330 Cham, Switzerland

Foreword

We live in extraordinary times. Examining the role of media, particularly digital platforms, has never been more important in understanding the new sociopolitical landscape in which environmental claims-making and decision-making take place. Social movements are simultaneously local and global (Castells 2012). At the click of a button connections can now be forged with like-minded individuals across the globe. Through mass action citizens can, on occasion, bring corporations to account. Environmental NGOs are able to bypass mainstream media by utilising the internet and social media to frame their messages on their own terms. Crowdsourced data, often using mobile phone applications, can provide a valuable source of on-the-ground knowledge and challenge official information. Social media are an important vehicle for mobilising individuals, connecting stakeholders, and raising income. There are also examples where individual solitary activism has spawned a huge following on social media, such as teenager Greta Thunberg—the new poster child of the environmental movement. These new tools have aided the growth of transnational communities of concern.

At the same time there has been a growing alliance between global NGOs and multinational corporations, operating across platforms no longer invested for the long term in the local, and supported by national and international governance regimes (Lester 2019). The 'local' for these global corporations and NGOs is 'locals', a succession of places and communities that supply resources or landscapes or the labour to harness them, and/or oppose corporate or government attempts to exploit their resources and lands for the global trade of goods, resources and people in which we are now all embedded. Yet, these 'locals' are also connected to each other in ways that are often surprisingly intimate. Communication media bring in contact local communities who rely on tourism on the Great Barrier Reef to villages in India struggling with the impact of coal burning on their air quality and the health of their children. They form relationships based on shared concerns that often bypass these transnational giants.

This new edited collection seeks to understand some of the major shifts that are shaping the media landscape and examines the implications for research agendas in the environmental communication field. As we become more and more connected across the globe through digital technology, news about our geographical locality reinforces our connection to place (Hess and Waller 2017; Howarth and Anderson 2019). Recent years have also seen the increasing emergence of 'hyperlocal' journalism—online ventures often run by citizen journalists that are geographically based and community oriented and tend to focus on neighbourhood issues such as the protection of a local green space (Williams et al. 2015).

Considerable optimism accompanied the rise in digital platforms and practices, including social media, with hopes that they would foster a new participatory dialogue and free exchange of ideas, but the old power struggles persist. The rise of right-wing populism has become one of the defining features of politics since the global recession in 2009. Populist movements have been on the rise for a number of years but reached a new point in 2016 with the election of Donald Trump in the USA and the Brexit referendum in the UK. Right-wing populist parties promote socially conservative and nationalist values, drawing on deep attachments to heritage and place, and they tend to express hostility towards policy addressing climate change (Lockwood 2019). While scientific evidence about global environmental issues such as climate change is frequently dismissed by right-wing populists as politically biased, national landscapes (countryside and coast) are given considerable symbolic value within their rhetoric. In an increasingly connected and, at the same time, disconnected world an appeal to people's roots is very powerful.

Digital media have provided a major platform for the spread of populism, but it is important to acknowledge that legacy media and social media are hybrid media systems that feed off each other (Chadwick 2017; Postill 2018). Nevertheless, social media have played an increasingly central role in the spread of propaganda and fake news, especially among audiences with low levels of literacy and/or media literacy (Cohen 2019).

The architecture of social media algorithms tends to lead to online filter bubbles, echo chambers and homophily, encouraging individuals to inhabit a hyper-partisan view of the world (Anderson 2014). Meanwhile, local journalists and environmental defenders who seek to make visible to wider audiences the illegalities and corruption driving environmental destruction have increasingly been targeted with imprisonment and violence (Reporters without Borders 2015; Lester 2017; Global Witness 2018).

Developments in Big Data also present a double-edged sword. While it has brought about enormous benefits in the collection and visualisation of huge volumes of data on environmental issues, it is increasingly colonising our lives with the growth of personalised marketing. Personal data may now be appropriated for a variety of ends, which has implications for environmental activists' use of digital media (Couldry and Mejias 2018, 2019). In a mediatised society, tracking is a constant feature of life, which raises concerns about continuous automated surveillance. In the UK, for example, social media surveillance of individual campaigners engaged in anti-fracking activism has been used to strengthen court cases against them (Hope and Collett-White 2018).

A key struggle of our time is between using platforms to make money and extract information, and using them to reinvigorate local communities and lives. What, we should ask, is the potential of media to help us live 'good' lives, where a 'good life' means having access to those places and things that support and enrich our experience, supplying basic necessities and those comforts that sometimes bring joy, and providing us with a political voice that allows us to influence the decisions that are made about our shared futures?

Traditional local media have experienced financial problems in sustaining themselves as they compete in an increasingly crowded marketplace. Ownership is less likely to be local and journalists are themselves less likely to be rooted in a particular place for an extensive period of time (Hess 2012). What we mean by 'local' and 'community' is contested and not straightforward to define in a globalised world. As Hess and Waller (2018: 37) observe: "Too often the local and community are positioned as something tangible, synonymous with a town or city – and serve as the structure in which journalism is examined. … the concept of community is not natural or inevitable. It is constructed by a series of discourses about society." Despite the unceasing synergy between geography and news, it would be wrong to assume that a local news outlet just serves a particular geographic area or is completely produced there (Hess and Waller 2018).

Indeed, the local matters. And research suggests that people are often more attached to their geographic communities than to online communities (Mersey 2009). This new volume illustrates how the expansion of digital networks and the capacity to communicate our local places have transformed the ways in which places are known and understood, adding a global dimension to the ways in which the 'local' is experienced.

We share many of the questions raised by chapters in this volume. We are a long way, for example, from fully understanding—let alone integrating—all that indigenous communities have to teach us about how to communicate a relationship with place, with land and the lives they support. And the grief and destruction to lives that occur when the land is lost and the relationship to place is severed. And what of those of us complicit in land theft through colonisation—that is, a large part of the world's population? It is vital that we learn to live with care, respect and love on stolen land, while acknowledging the rights and the knowledge about the land that exists in indigenous communities. Importantly, we still have much to learn about how that knowledge was communicated from generation to generation and along vast networks of trade and culture.

Nor do we yet know what role science will assume in the future, particularly how it might contribute to resolving tensions over local resources and landscapes, and the impacts on local communities. While science played a paramount role in what authors describe as a first wave of contemporary environmental communications (Earth Rise images and *Silent Spring* are key examples—see chapter "Environmental Communication in the Intertwining of the Local and the Digital"), science has never played a simple, uncontested role. In 1962, the silence of the birds in a symbolic representation of the 'local' was so powerful that chemical companies and the science that continued to support them marshalled forces to diminish Rachel Carson's findings and the scientist herself (Hazlett 2004). Faith in 'the science' can be just as alienating as faith in 'economic growth' or 'man's God-given rights over nature'.

The global became a defining motif of a second wave of contemporary environmental communications. For scholars, the imperative was to understand how global capital, risks and communication networks combined to impact our shared environmental futures. Divisions, such as 'the West' and 'Global South-Global North', helped us analyse the world at scale. For a third wave, we are developing new tools to understand digital communications, detailed approaches to analysing the local, and more nuanced categories in which to view the world. This is because we recognise, as this volume illustrates so strongly, that the local is where change is experienced, where loss is articulated, and where resistance begins.

Plymouth, UK Alison Anderson
Hobart, Australia Libby Lester

References

Anderson, A. (2014). *Media, environment and the network society*. London: Palgrave Macmillan.

Castells, M. (2012). *Networks of outrage and hope: Social movements in the internet age*. Cambridge: Polity.

Chadwick, A. (2017). *The hybrid media system: Politics and power*. Oxford: Oxford University Press.

Cohen, A. C. T. S. (2019). Exercise of citizenship in network society: Cyberactivism and the fight against fake news in the Brazilian context. In A. Melro & L. Oliveira (Eds.), *Constitutional knowledge and its impact on citizenship exercise in a networked society*. Information Science Reference.

Couldry, N., & Mejias, U. (2018). Data colonialism: Rethinking big data's relation to the contemporary subject. *Television and New Media*, *20*(4), 336–349.

Couldry, N., & Mejias, U. (2019). *The costs of connection: How data is colonizing human life and appropriating it for capitalism*. Stanford: Stanford University Press.

Global Witness. (2018). At what cost? Irresponsible business and the murder of land and environmental defenders in 2017. Retrieved from https://www.globalwitness.org/en/campaigns/environmental-activists/defenders-annualreport/

Hazlett, M. (2004, October). 'Woman vs. man vs. bugs': Gender and popular ecology in early reactions to silent spring. *Environmental History*, 701–729.

Hess, K. (2012). Breaking boundaries: Recasting the "local" newspaper as "geo-social" news in a digital landscape. *Digital Journalism*, *1*(1), 48–63.

Hess, K., & Waller, L. (2017). *Local journalism in a digital world: Theory and practice in the digital age*. London: Palgrave Macmillan.

Hope, M., & Collett-White, R. (2018). Investigation: Facebook surveillance of fracking protesters—Background and key documents. DeSmog, UK. https://www.desmog.co.uk/2018/06/15/investigation-fracking-company-facebook-surveillance-environmental-protesters-key-information-and-documents

Howarth, C., & Anderson, A. (2019). Increasing local salience of climate change: The un-tapped impact of the media-science interface. *Environmental Communication*, *13*(6), 713–722.

Lester, L. (2017). Environment and human rights activism, journalism and the "New War". In S. Waisbord, & H. Tumber (Eds.), *Routledge companion to media & human rights* (pp. 268–276). Abingdon/New York: Routledge.

Lester, L. (2019). *Global trade and mediatised environmental protest: The view from here*. Basingstoke: Palgrave.

Lockwood, M. (2019). Right-wing populism and climate change policy. https://www.oxfordresearchgroup.org.uk/blog/right-wing-populism-and-climate-change-policy

Mersey, R. D. (2009). Online news users sense of community. *Journalism Practice*, *3*, 347–360.

Postill, J. (2018). Populism and social media: A global perspective. *Media, Culture & Society*, *40*(5), 754–765.

Reporters Without Borders. (2015). Hostile climate for environmental journalists. Retrieved from https://rsf.org/sites/default/files/rapport_environnement_en.pdf

Williams, A., Harte, D., & Turner, J. (2015). The value of UK hyperlocal community news. *Digital Journalism*, *3*(5), 680–703.

Acknowledgements

The work with this volume began in conversations between co-editors at the International Association for Media and Communication—IAMCR Conference 2017 in Cartagena, Colombia. The co-editors are variously involved in leading environmental communication themes and sections in IAMCR and ECREA (European Communication Research and Education Association). We thus acknowledge these international media and communication organisations, IAMCR and ECREA, which have provided the arena for us to meet and exchange ideas. Alongside the volume proper has been the global exchange of ideas, greater understanding of similarity and difference, and personal and professional development that has enabled a truly enriching scholarly experience.

One might imagine that bringing such global diversity to a single volume has its challenges other than arranging regular Skype meetings across time zones. The co-editors span the globe—residing in Spain, Belgium, Sweden, India and Australia—and have come to environmental communication from often different perspectives, bringing their local experience and scholarship to meet with, and at times, collide with, others. We, as co-editors, have commented regularly on the spirit of collaboration, kindness and respect that has been exhibited throughout the process of bringing this volume together. We celebrate how we have navigated the challenges implicit in such a task and thank each other for our friendship and collegiality—an output not readily seen in our table of contents but heartfelt and treasured.

As co-editors, we have also had the privilege of working closely with individual authors, guiding their work to publication. We thank our authors for their work in producing their chapters, for their intellect, efficiency and patience. Like the co-editors, the chapter authors and their environments span the globe and, when brought together, showcase the vibrant, diverse and innovative field that characterises environmental communication research. We are honoured that eminent scholars in the field Alison Anderson and Libby Lester provided the foreword for this volume. Furthermore, we thank Lucy Batrouney, Mala Sanghera-Warren and Bryony Burns at Palgrave Macmillan for their carriage and publication of this collection. And a special thanks goes out to the Palgrave/IAMCR book series editors Marjan de Bruin and Claudia Padovani. We are grateful to our various institutions that support our work in environmental communication and our mentors that help guide us through the world of academia.

We thank our families and loved ones for their continuous support.

And finally, we acknowledge our planet and all nature that gives us life. Our deepest hope is that in the intertwining of the local and the global, we might find ways to better care for our earth and each other.

About the Book

The Birth of an Idea

The seeds for this volume were planted during the 2017 conference of the International Association for Media and Communication Research (IAMCR) in Cartagena, Colombia, from 16 to 20 July. A new management committee had been elected for the Environment, Science and Risk Communication Working Group (ESR WG) at the previous conference in 2016 in Leicester: Pieter Maeseele (University of Antwerp, Belgium) was re-elected as Chair, and he was joined by the newly elected Vice-Chairs Joana Díaz-Pont (Autonomous University of Barcelona, Spain), Kerrie Foxwell-Norton (Griffith University, Australia) and Maitreyee Mishra (Manipal University, India). Anders Hansen (University of Leicester, UK), who had founded the Working Group at the 1988 IAMCR conference in Stockholm, and Dorothee Arlt (University of Bern, Switzerland), who had been involved in the ESR WG since 2013, had both stepped down as Vice-Chairs.

It was after the business meeting of the ESR WG that the idea was first contemplated between the both of us to compile an edited volume based on work presented at the ESR WG sessions. We had been struck by how many papers emphasised the importance of place in environmental communication (research). This seemed like an important turning point for the field, after about two decades in which the global had been emphasised over the local within the context of an undisputed reign of a discourse of globalisation. A second observation was the centrality of the role and performance of digital communication platforms in many presentations, clearly superseding a traditional focus on print in environmental communication research. It was a sign of the times as we had been witnessing both the election of Donald Trump as US president and the 'Brexit vote' in the UK European Union membership referendum vote the year before. Both of these events had been characterised as heralding the return of values such as self-determination and popular sovereignty, or nationalism and nativism, in the highest levels of government in two leading nations of the world system (i.e. 'the local') and as ushering in a new and substantial role for digital communication platforms such as Facebook in elections or referendums (i.e. 'the digital').

From the start, it was obvious that this volume would be a joint initiative of IAMCR's ESR WG, on the one hand, and the Science and Environment Communication Section of the European Communication Research and Education Association (SEC ECREA), on the other. Kerrie and Maitreyee were already doing important work on the role of place in environmental communication, with their focus on indigenous communities in Australia and India, respectively. Secondly, Pieter was also vice-chair of SEC ECREA at the time, and under the leadership of its chair Annika Egan Sjölander (Umeå University, Sweden), SEC ECREA was already preparing a symposium that would invite scholars to reflect on the existence and implications of a presumed 'digital turn' in science and environmental communication. In other words, the officers of both sections found a common interest in addressing the need for a scholarly debate on how digitalisation and the changing role of place affect environmental communication as a research field. The idea was pitched to Marjan de Bruin and Claudia Padovani, the co-editors of the Palgrave/IAMCR books series —Global Transformations in Media and Communication Research. A call for papers was drafted thereafter and widely disseminated in the fall of 2017.

The Call for Papers

With the tentative title "When the Local Meets the Digital—Implications and Consequences for Environmental Communication", the call focused on how recent changes in territorial and digital capabilities of communication pose new challenges for environmental communication, with particular impacts and consequences at the local level. It emphasised how a redefinition of environmental problems was called for in the context of a shift from the well-known maxim "think globally and act locally" to the new "think locally and act locally". At the same time, it emphasised how the local impacts of social media and the emergence of collaborative platforms were exposing multifaceted realities, drawing on the example of how the discourses on the diverse benefits of platforms such as Airbnb, Uber or Amazon or new developments such as SmartCities had been found to contrast with their physical and social consequences on the ground.

The call for papers explicitly started from a critical perspective, integrating concerns from both critical political economy and cultural studies. It observed how the intertwining of place and digitalisation produces new social and environmental injustices which, in turn, could be expected to trigger the emergence of new social movements and citizen engagement to fight the effects of decisions taken at distant territorial levels and in digital spaces disconnected from the local—especially as these distant decisions often support profits and extractivist interests that are the targets of local protest. In other words, it called for a recognition and exploration of how social relations at a local level are impacted by the integration of global communication platforms in the daily lives of citizens and cities. The questions that we put forward that could potentially guide the contributions were the following:

- How is environmental communication impacted by the intertwining of the local and the digital?
- How are digital media influencing environmental communication at the local level?
- What is/could be the role of environmental communication in the emergence of local environmental activism, networking, and political and social participation?
- What is/could be the role of journalism and news media—mainstream and/or alternative—in navigating the local and digital?
- What are the local experiences of digital collaborative platforms (e.g. Airbnb, Uber or Amazon) that, in so many instances, have become extractive platforms?
- How, if at all, are environmental or social movements created or redefined in response to these new injustices originating in distant digital spaces?
- How does the prevailing political economic system impact on the local experience of 'Green Growth' initiatives, such as SmartCities?

The Catch

By February 2018, 19 extended abstracts had come in. The authors of 16 proposals were invited to present their proposals during an SEC ECREA webinar on 15 March 2018. At this travel-free meeting these proposals were discussed with an audience of about 45 participants. Fifteen draft chapters were submitted by June, after which nine were eventually accepted after a lengthy reviewing process in the latter half of 2018. Furthermore, two more chapters were commissioned from authors working on cases related to Africa and Brazil, in response to an encouragement of the reviewers of our manuscript proposal to Palgrave to seek additional contributions from the southern hemisphere. And last but not least, Alison Anderson (University of Plymouth, UK) and Libby Lester (University of Tasmania, Australia) accepted an invitation to write a foreword to the volume, for which we are very grateful.

There were many interesting differences between the original call for papers and the eventual contributions that were received. Firstly, there were no proposals discussing the role of the digital platforms Airbnb, Uber or Amazon, while Facebook and WhatsApp played a central role in many proposals, in addition to georeferencing apps and spatial media. Secondly, hardly any proposal addressed the role of journalism and news media—either mainstream or alternative—in navigating the local and the digital, which is an obvious deviation from traditional environmental communication research. Thirdly, the digital dimension was clearly the starting point for many contributions, while the role of place often appeared to be a secondary concern. Just a few proposals emphasised the changing role of place over digitalisation. Fourthly, the proposals seemed to respond particularly to the following two, closely related, questions from the original call: What is/could be the role of environmental communication in the emergence of local environmental activism, networking, and political and social participation? How, if at all, are environmental or social movements created or redefined in response to these new injustices originating in distant digital spaces? And last, several proposals came in that discussed the instrumental use of digital tools in, for example, disaster and risk communication, instead of reflecting on how the intertwining of place and digitalisation impacts changes in the social construction of the environment.

At the same time, what was remarkable about the eventual submissions was their large diversity in terms of environmental concerns and issues, forms of media and communication practices, and analytical approaches and methodologies—and this, moreover, from a very wide geographical range of scholars, subjects, areas and locations. It allowed us to present an empirical case study collection from different parts of the world in which the interaction between local and place-based environmental concerns and problems, on the one hand, and digital media technologies and practices, on the other, is investigated in each chapter. Because most chapters discuss recent cases from different countries in which environmental activists and campaigners creatively use digital technologies for campaign purposes, the volume identifies new environmental communication challenges and opportunities for, as well as practices of, environmental activists, campaigners, NGOs and local communities, while discussing their implications for environmental communication (research).

The volume includes a large diversity of environmental issues and concerns (conservation, air pollution, renewable energy, farming, water management, land rights, food governance, home renovation, etc.), forms of media and communication (drones, video streaming, mobile apps, Facebook, Instagram Twitter, etc.), and analytical approaches (political economy, culturalism, organisational communication, discourse analysis, social construction of technology, etc.) and methodologies (ethnography, interviews, text analysis, surveys, focus groups, software design, etc.). And this, moreover, from a wide geographical range of scholars, subjects, areas and locations: while the editors of the volume are based in three different regions of the world (Europe, East Asia, and Oceania), the list of authors adds two more large regions to this list (North and South America). And the chapters themselves deal with even greater geographical diversity, as also sub-Saharan African countries and contexts are present in the case studies. On the other hand, it is important to acknowledge that we have failed to include work from North and Central Asia, the Middle East, and the Pacific Islands, and that Africa and South America clearly remain under-represented.

This edited collection brings into focus how local and digital spaces influence environmental communication activities in various domains with differing outcomes. We hope the different case studies about how social networking apps and other digital technologies *are* effectively used by environmental activists, workers and social movements in different parts of the world—and how they *can be* effectively used in the future—will inspire you in the fight for social and environmental justice.

Pieter Maeseele
Joana Díaz-Pont

Contents

Notes on Contributors

Deb Anderson is an academic and journalist based in Melbourne, Australia. After a decade of working with Fairfax quality newspapers *The Sydney Morning Herald* and *The Age*, she joined Monash University as a lecturer in 2013. Her research explores the experience and meaning of extreme weather events for ordinary people, in a time of politicised discourse on climate. She is also examining the professional and ethical challenges of reporting weather disasters, funded by the Journalism Education and Research Association of Australia. Her works have been published widely, and she is the author of *Endurance: Australian Stories of Drought* (2014).

Claudio Broitman is communication sciences' master director in the School of Journalism at the University of Santiago (Chile). He obtained his PhD at Sorbonne University (Paris IV) in 2017 and his master's degree at École Normale Supérieure de Lyon (2016). He researches in the science studies field, especially on science communication, knowledge production and environmental communication. His main publications are in Minerva and in Questions de communication.

Joana Díaz-Pont is a lecturer and researcher in the Department of Political Science and Public Law at the Autonomous University of Barcelona. Her main research interests are environmental communication, governance and policies. As a practitioner she collaborates with international and environmental agencies and corporations, and one of her latest publications explores the relationship between corporate social responsibility and corporate governance (Emerald, 2017). She is the vice-chair of IAMCR's Environment, Science and Risk Communication Working Group.

Annika Egan Sjölander is Associate Professor of Media and Communication Studies at Umeå University, Sweden. Communication for sustainable societies in the Anthropocene summarises her main research interest, evolving around the role of media and communication and the handling of pressing environmental problems such as climate change, nuclear waste and water scarcity. Egan Sjölander is a founding member of the International Environmental Communication Association (IECA) and has been Chair of the Science and Environment Section in ECREA (2012–2018). Her discourse analytical interest resulted in the co-edited volume *Tracking Discourses. Politics, Identity and Social Change* (2011).

Camila Emboava Lopes is a PhD candidate in the Department of Culture and Media Studies at Umeå University, Sweden. Her research project is about identity construction, articulation and negotiation in and through indigenous-owned media initiatives in contemporary Brazil. Her main research interests are global and local indigenous movements, digital media, social media and protest. She has a master's degree in indigenous journalism from the Sámi University of Applied Sciences in Guovdageaidnu/Kautokeino, Norway, and a bachelor's degree in journalism from the Federal University of Mato Grosso do Sul in Campo Grande, Brazil.

Kerrie Foxwell-Norton is Associate Professor of Communication, Media and Journalism Studies at Griffith University (Australia). Her research explores environmental communication of coastal and marine environments and the myriad of communities therein. She is the author of *Environmental Communication and Critical Coastal Policy: Communities, Culture and Nature* (2018).

Anders Horsbøl is an associate professor in the Department of Communication and Psychology at Aalborg University, Denmark. With a background in philosophy, communication studies and discourse analysis, he has explored approaches to and dilemmas of dialogue, participation and democracy within different societal realms. His works have been published particularly in the fields of (multimodal) discourse analysis, political communication, health communication, citizen engagement and environmental communication.

René Jara is a journalist, a graduate in communication from the University of Chile and a doctor of political science from the Universidad de Grenoble-Alpes. He is the director of the School of Journalism at the University of Santiago (Chile). He has worked as a professor in Chilean and French universities, in the fields of communication and political science. His research lines are in the crossroads of the fields of communication, socio-history and political science. He conducts a research team, funded by a baseline project (MECESUP USA1555), the purpose of which is to apply computerised tools, data analysis and graphos theory to the understanding of social and political problems.

Yowei Kang is an assistant professor at the Bachelor Degree Program in Oceanic Cultural Creative Design Industries, National Taiwan Ocean University, Taiwan. His research interests focus on new media design, digital game research, visual communication, and experiential rhetoric. Some of his works have been published in *International Journal of Strategic Communication* and *Journal of Intercultural Communication Studies*. He has received government funding to support his research in location-based advertising and consumer privacy management strategies.

Pieter Maeseele is an associate professor in the Department of Communication Studies at the University of Antwerp (Belgium). His research agenda has evolved from a focus on social and political issues relating to science, technology and the environment in media to a broader focus on the relation between media and democracy. He is the chair of IAMCR's Science, Environment and Risk Communication Working Group, past vice-chair of ECREA's Science and Environment Communication Section and a member of the editorial board of Science Communication and the Palgrave Studies in Media and Environmental Communication.

Alana Mann is the chair of the Department of Media and Communications, Faculty of Arts and Social Sciences (FASS), University of Sydney, Australia, and a key researcher in the Sydney Environment Institute. Her research focuses on the communicative dimensions of citizen engagement, participation and collective action in food systems planning and governance. She is the author of *Voice and Participation in Global Food Politics* (2019) and a chief investigator on the Australian Research Council (ARC)–funded project FoodLab Sydney.

Maitreyee Mishra is an assistant professor at Manipal Institute of Communication, MAHE, India, where she has been teaching since 2010. She does research across environmental humanities, development and gender, Japanese anime and anthropology, and her works have been published in these fields. Her recent work on Indian environmental journalism will appear in the *Routledge Handbook of Environmental Journalism* (forthcoming).

Derek Moscato is Assistant Professor of Journalism and Public Relations at Western Washington University. His research focuses on the intersection of environmentalism, persuasion and international communication.

Patrick D. Murphy is Associate Dean for Research and Graduate Studies and associate professor in the Department of Media Studies and Production at the Klein College of Media and Communication, Temple University. His research interests include global media, environmental communication and ethnographic method. Murphy is the author of *The Media Commons: Globalization and Environmental Discourses* (2017) and co-editor of *Negotiating Democracy: Media Transformation in Emerging Democracies* (2007) and *Global Media Studies* (2003). His work has also appeared in numerous journals and as chapters in edited books.

Aneta Podkalicka is a researcher and Lecturer in Communications and Media Studies at Monash University, Melbourne, Australia, dedicated to collaborative, interdisciplinary academic work that engages communities and industry. She has conducted research in the areas of social inclusion, consumption, and economic and environmental sustainability. She is the co-author of *Using Media for Social Innovation* (with Ellie Rennie, 2018) and *Grand Designs: Consumer Markets and Home-Making* (with Esther Milne and Jenny Kennedy, Palgrave 2018).

Emmanuel Septime Sessou is a doctoral student in the Media and Communication doctoral programme at Temple University's Lew Klein College of Media and Communication. His research interests are within the scope of geopolitics, development communication and environmental communication in Africa.

Matteo Tarantino is an assistant professor in the Dipartimento di Scienze della Comunicazione e dello Spettacolo at the Università Cattolica di Milano and a lecturer at the University of Geneva's Institute for Environmental Sciences. His works on the topics of data, cities and sustainability, especially in the Chinese context, have been published extensively. He co-founded the Media and the City working group within ECREA. He collaborates with UN Environment and the UN Economic Commission for Europe.

Tracy M. Tinga is Assistant Professor of Global/Transnational Media at the University of Maryland, Baltimore County. She received her PhD in media and communication at Temple University's Lew Klein College of Media and Communication. Her research interests focus on media and globalisation, environmental communication, and communication for development and social change as they relate to Africa.

Franzisca Weder is a senior lecturer at the School of Communication and Arts at the University of Queensland, Brisbane/Australia and associate professor at the University of Klagenfurt, Austria (on leave). Her research extends across strategic communication, public relations and organisational communication, with a strong focus on sustainability, Corporate Social Responsibility (CSR) and environmental communication. She is the vice-chair of the International Environmental Communication Association (IECA) and the chair of the Austrian Communication Association (ÖGK). Her publications include her recent work on CSR Communication (CCIJ, 2019), based on two books on CSR from an organisational communication perspective (2008, 2012), and journal papers on storytelling as innovative methodology (*Sustainability*, 2019).

Kenneth C. C. Yang is a professor in the Department of Communication at the University of Texas at El Paso, USA. His research focuses on new media advertising, consumer behaviour, and international advertising. Some of his many works have been published in *Cyberpsychology*, *Journal of Strategic Communication*, *International Journal of Consumer Marketing*, *Journal of Intercultural Communication Studies*, *Journal of Marketing Communication*, and *Telematics and Informatics*. He has edited or co-edited three books: *Asia.com: Asia Encounters the Internet* (2003), *Multi-Platform Advertising Strategies in the Global Marketplace* (2018) and *Cases on Immersive Virtual Reality Techniques* (2019).

Chamee Yang is a PhD researcher at the Institute of Communications Research (ICR), the University of Illinois at Urbana–Champaign. Her dissertation "Remapping Songdo: A Genealogy of a Smart City in South Korea" investigates South Korea's modern history of urbanisation, militarisation, digitalisation and globalisation manifested through the multiple problems of the smart city—mobility, security, environment and futurity—in the twenty-first century.

List of Figures and Table

Old Power Struggles and New Media Work: Indigenous Peoples' Striving for Justice in Contemporary Brazil

Sparking an Interest in Local Water Issues: Social Media as an Enabler for (Hyper)local Community Engagement in Environmental Issues and the Role of Solution Journalism

Environmental Communication in the Intertwining of the Local and the Digital

Joana Díaz-Pont, Annika Egan Sjölander, Kerrie Foxwell-Norton, Maitreyee Mishra, and Pieter Maeseele

Introduction

In February 2019, the world's news media reported dozens of polar bears invading a small community in the Russian archipelago of Novaya Zemlya in the Arctic Ocean. From *The Siberian Times* to *The Washington Post* to *El País* to *The Guardian Australia* to India's *The Hindu*, the story went viral. It appeared on our television sets, in our Instagram feed, and as news

J. Díaz-Pont (✉)
Department of Political Science and Public Law, Autonomous University of Barcelona, Bellaterra, Spain
e-mail: Joana.diaz@uab.cat

A. Egan Sjölander
Department of Culture and Media Studies, Umeå University, Umeå, Sweden
e-mail: annika.egan.sjolander@umu.se

J. Díaz-Pont et al. (eds.), *The Local and the Digital in Environmental Communication*, Global Transformations in Media and Communication Research - A Palgrave and IAMCR Series,
https://doi.org/10.1007/978-3-030-37330-6_1

on our Facebook sites. The story was tweeted and shared innumerable times throughout the network of networks that crisscross the earth and its atmosphere, variously bringing the news of this small place to global audiences. This moment of environmental and, more specifically, climate change communication—melting Arctic ice sheets, the polar bears' principal home, has caused them to spend more time on land, seeking food—epitomizes much of the challenges in the intertwining of the local (and the global) in digital networks. Beyond its mere mediation of climate change impacts lie many more complex issues: geopolitics and the unequal distribution of climate change costs, risks and impacts; distant and sparse populations of the Arctic region and its contrast with the dense populations of urban centres more culpable; the 'spectacular' of polar bears and both their literal and their figurative distance from, for example, the animals of industrial agricultural production and so on. In Novaya Zemlya, we confront again the core challenge of environmental communication in an era of ecological crises and devastation: how can we communicate these local experiences, across diverse contexts, to foster responses and action on global environmental issues? An examination of the intertwining of local experiences and digital networks—the contemporary conditions of environmental communication—provides some direction. These linked spatial and temporal networks communicate the social construction of the environment, working to both enable and limit our responses. How we understand the proverbial polar bears and their relationship to climate change and humans is constructed in these networks of environmental communication.

K. Foxwell-Norton
School of Humanities, Languages and Social Science, Griffith University, Gold Coast, QLD, Australia
e-mail: K.Foxwell@griffith.edu.au

M. Mishra
Manipal Institute of Communication, Manipal Academy of Higher Education (Manipal University), Manipal, India
e-mail: maitreyee.mishra@manipal.edu

P. Maeseele
Department of Communication Studies, University of Antwerp, Antwerp, Belgium
e-mail: pieter.maeseele@uantwerpen.be

Environmental communication scholarship has drawn attention to the social construction of our environments (Anderson 1997; Cox 2006; Lester 2010). The contexts and cultures of environmental communication, constituted in ideologies and discourses that define histories and places, also define its rhetoric, framing and form as symbolic action (Burke 1966; Carvalho 2007). In the process of understanding the variables that define environmental problems, we learn more about environmental communication and its challenges. In this chapter we examine environmental communication by focusing on two variables: the *geographical*, understood here as places global as well as local where the environment is appropriated by social actors; and the *digital*, as a means to explore social interconnectivity enabled by different forms of communication technology that in turn can strengthen networks and social action. We take a retrospective look at environmental communication to highlight how the geographical and the digital entwine.

If we characterize environmental communication according to how it has dealt with geographical scale and social interconnectivity in the last 60 years, we could probably affirm that we are now entering what can be described as a third wave. The first wave started in the 1960s as a response to environmental and human disasters that had a great impact on both a local and a regional scale. Those events echoed loudly across the globe as scientists and NGOs partnered to protest the nefarious impacts of industrial production at a global scale. For the first time, environmental communication went global and actors made prominent use of audiovisual media to raise awareness and place the environment on political agenda. During this first wave, environmental issues relied on scientists to gain impact on the agenda, and on activists and the audiovisual media system to be able to reach out globally (Bell 1994; Hansen 1993). The famous 'Earthrise' photo taken from space by the *Apollo 8* Mission in 1968 and the release of Rachel Carson's *The Silent Spring* in 1962 are but two examples that point to this first wave, where the impacts on our fragile, beautiful planet were communicated to the public via media that was at once local and global in its reach.

A second wave started in the mid-1990s as the internet developed and partially replaced traditional (mass) media. Intense global debates triggered by the pursuit of Environmentally Sustainable Development (ESD) took place to balance national economies with environmental protection policies. By the 1990s, ESD was being embraced enthusiastically by national governments, especially evident at the 1992 Rio Earth Summit and in the publication of *Agenda 21* (1992). The internet eventually enhanced transnational interconnectivity and helped global issues such as biodiversity loss and climate change to be communicated in society at large, in politics and across different markets (Castells 2009). In that process, new actors came to dominate global environmental communication. These were actors who were already leading the globalization debate, namely, on the one hand, nation-states that negotiated environmental agreements on desertification, biodiversity, climate change and so on; and on the other, transnational corporations that adhered to global ethical frameworks epitomized by the emergence of corporate social responsibility. They had in common that these newly formulated 'green' efforts were communicated via their relatively new institutional websites (Díaz-Pont 2000; Wood and Logsdon 2001). International NGOs watched and used the same World Wide Web to denounce on a global scale the abuses committed in different regions of the world (Kilbourne 2004), using their own media to expose or challenge governments' and industries' environmental practices.

A third wave in environmental communication is emerging and is mediated by social networks both digital and place-based. The actors have changed to some extent and so have their political nature, their power and their capacity to interact. In the network society, the environment faces new challenges in a rapidly changing scenario for both media and environmental activism (Anderson 2014). The geographical scale has also changed and so have the claims as they are now more often linked to the territory and its sovereignty in the environmental political and economic agenda (Jessop 2016; Swyngedouw 2017). Social media have allowed, if not always, at least in specific contexts, a progressive deinstitutionalization of environmental communication. One such example is the Skolstrejk för klimatet (Fridays For Future) movement initiated by the Swedish teen Greta Thunberg. Different from previous movements led by NGOs or more or less coordinated platforms, this is an example of solitary activism that evolves and mobilizes followers on social media platforms. These platforms are providing local actors with tools and strategies to actively participate in the definition and resolution of their environmental agenda. In other words, local activists and community advocates using digital networks are trying to regain the debate in search of solutions that are people- and place-based. Some of these solutions are evident in the development of 'sharing economy' (i.e. peer-to-peer sharing of goods and services) and different forms of collaborative platforms, with as yet unclear impacts on sustainability transitions (Martin 2016).

One can wonder why should the timeline of environmental communication be framed as a sequence of waves? And why should we emphasize the last 60 years of environmental communication? Our own debates as co-authors from diverse cultural backgrounds and intellectual traditions have revealed the nuances and different perspectives that can motivate many questions concerning the wave approach. We acknowledge the weaknesses, and our intention is not to make firm claims that these divisions are the only possible narratives at hand. We only prefer it here as an approach that captures significant features of a certain time period and allows us to compare different historical transformations. The features of each wave help us to frame the subject of this volume as well as the object for analysis: how the local and the digital matter for environmental communication.

In our attempt to characterize phases, we assume that transitions from one stage to another are not results of sudden changes, or political, technological or cultural revolutions. Our argument suggests that these transitions signal the dialectics between geographical scale and social interconnectivity. This third wave is particularly characterized by the intertwining of the local and the digital, reformulating the old but persisting debates on power, wealth and position. This period builds upon established research in the field of environmental communication, especially studies concerned with representation in the form of the construction of messages, frames and discourses; but widens to embrace and reformulate debates on democracy, governance and other themes of political inquiry, especially in relation to climate change (Newell et al. 2015; Nisbet et al. 2014; Maeseele 2015).

In this introductory chapter, we explore the intertwining of the local and the digital in a changing landscape for environmental communication that is also being analysed in other environmental disciplines (Stevenson and Dryzek 2013; Swyngedouw 2017). We examine these changes in four steps. First, we present the three waves in a 60-year timeline of environmental communication, focusing on how geographical scale and social interconnectivity have changed in each phase and how these changes have affected the discipline. Second, we focus on place, and the implications of a reconfiguration from a global to a local outlook in environmental communication, on the one hand, and intensified connections between place and global movements, on the other. Third, we look at how recent research has paid attention to how digitalization, and networks and social media platforms in particular, are transforming environmental communication. We conclude that the intertwining of the local and the digital poses questions for environmental communication that remain unanswered. Simultaneously, it is obvious that the pursuit of an enhanced understanding of its implications should be at the fore of our scholarship. The different chapters that we present in this volume address these questions and implications in various ways.

Three Waves in Environmental Communication

Environmental communication has a much longer history than the naming of a discipline might indicate. Within the Western academic world, predominantly the US, it can be observed that in the 1960s, environmental communication emerged partly from being a subfield in scientific communication to deal with substantive foci of research that included messages, perception, framing and content. According to Durant et al. (1989), for at least the three following decades, communication of environmental issues was very much linked to disasters, scientific claims and the public understanding of issues. A recent literature review of the field of environmental communication scholarship similarly found origins of environmental communication scholarship in science communication and risk communication (Comfort and Park 2018). Like these authors, we are reluctant to define a single genesis of environmental communication scholarship. Uneven flows of scholarship favouring northern hemisphere definitions and the relative silencing of lesser volumes of scholarship that may have emerged from different traditions (i.e. not science and risk) are cautionary tales—alongside the persistent marginalization of First Nation's cultural heritage, so deeply steeped in 'environmental communication'.

With these caveats, to describe three waves of environmental communication we begin with the birth of modern environmental consciousness during the 1960s. Figure 1 represents this sequence of waves, although we need to clarify that the objective is not to be exhaustive, but only to highlight certain events that can help exemplify the transformations in nature of well-known milestones in environmental communication.

Some key characteristics of the first wave relate to the nature of communication actors and their aim to bring local and regional issues to the global agenda. Networks between local actors and their connections around the globe depended on social and scientific leaderships with

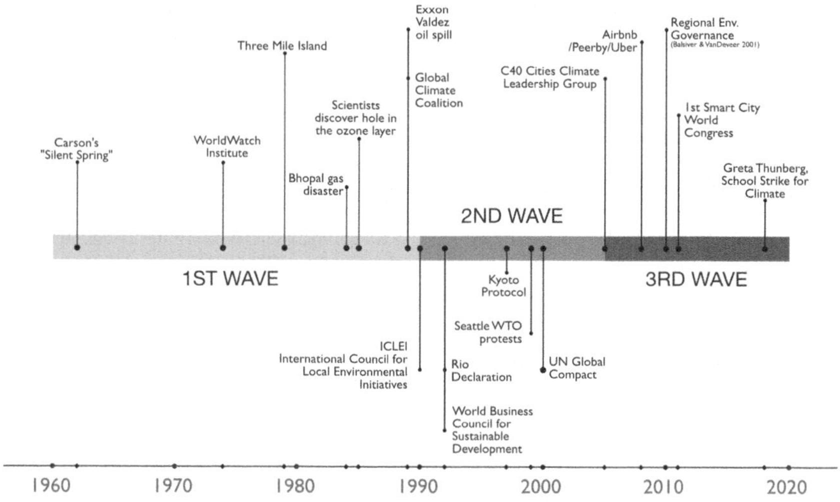

Fig. 1 A timeline of environmental communication. (Source: Own elaboration)

well-identified public actors leading debates that sought to capture global attention (Mazur and Lee 1993). Environmental communication during this period can be defined as unidirectional and global-oriented. The environmental movement in conjunction with the specialized scientific community learned together how to garner public attention at the global scale by involving mass media in the process of problem construction. During this first wave, environmental communication research focused primarily on message construction and therefore also on the role of journalists as mediators of scientific content. Their need for specialization was also stressed (Anderson 1997), alongside the constructivist approach to message-building (Mazur 1984) and meaning creation (Gamson and Modigliani 1989). The scale merely amplified political and economic dimensions, and goals were very much limited to awareness-raising and agenda-setting. Some of the key challenges regarding environmental communication in this period were choosing sources, covering news, depicting issues, creating meaning and stimulating forums. The media, NGOs and scientific forums were the central actors trying to reach a global audience (Catton and Dunlap 1978; Muir and Veenendall 1996).

The transition to a second wave begins with the capacity of environmental communication to foster public participation, decision-making and conflict resolution (Kunst and Witlox 1993). In this transition, embryonic networks spread and new networks germinated. For example, the ICLEI—Local Governments for Sustainability—started in 1989 when 35 local government leaders met to discuss the ozone layer. Founded in 1991, the ICLEI is now a global network with more than 1750 local and regional governments committed to generating political awareness on key themes in the area of sustainability. Similarly, the WBCSD (World Business Council for Sustainable Development) started in 1991 as a business voice in global conversations at the Rio Summit. In 1995 it became a network of 120 members, and today it is a global network that promotes sustainable development at local and regional levels. Communication became more multidirectional and also multimodal. Different regions and new actors took the lead. NGOs were better versed in the strategic use of media, and traditional media also learnt how to react to this instrumentalization while generating a rebound effect on the part of governments and corporations. Over a decade after the ICLEI's founding, the C40 Cities Climate Leadership Group emerged in 2005 as a network of megacities committed to sharing knowledge and action on climate change. Less than a decade after the founding of the WBCSD, the UN Global Compact became a global movement of sustainable companies and stakeholders promoting principles on human rights, labour, environment and anti-corruption (Chapman 1997; Deetz 1995; Hajer and Fischer 1999).

The global audience were no longer passive consumers of information by the midst of the first decade of 2000, compared to the static, more or less fixed role of the public during the first era of environmental communication. Environmental perceptions and values started to build during the first wave but became behaviours and claims in the second wave. Environmental activism was enhanced with the introduction of the internet, especially its possibility to mobilize and coordinate action among stakeholders at a transnational scale. The Seattle protest in 1999 became a symbol of a new communication order that multiplied individual capacity without the need of clearly identified leaderships (Clark and Themudo 2006). This change stimulated governments and corporations to counter and reframe arguments and claims to a larger extent, and to also engage in and initiate participatory communication strategies (Delicath et al. 2004; Sen et al. 2006). This moment in time opens a *belle époque* for new networks in the social, political and economic arenas, with movements, parties and organizations vindicating their own power and role in the environmental debate (Kamieniecki 1993). The emergence and growth of power and influence of sectoral and regional environmental networks epitomizes new forms of environmental communication during this second wave. Today, we know that the internet of the 1990s had a kind of window function for governments, corporations and social organizations in the sense that they were now able to communicate their environmental agendas directly to their stakeholders, *without* the mediation of journalists working within traditional mass media. Governments and industry implemented corporate social responsibility strategies and databases, information and public participation systems, and other communication tools (Díaz-Pont 2017). However, this happened often as before in a quite one-directional format (Lischinsky and Egan Sjölander 2014; Schultz et al. 2013). Nevertheless, hypertextuality was treasured on the internet and interconnectivity between environmental communication actors, new and old, at all levels partly began to replace traditional mediator functions of journalists and mass media.

What we see as a potential third wave in the field of environment communication is the formation of a scenario where interactivity increases, not least due to the rapid expansion of different forms of social media in the last decade. The number of active communication actors not only multiplies, but the actors themselves also redefine and disrupt. This ongoing development needs further examination in order to be better understood, but there are indications that communicating environments, cultures and political and economic systems is more complex than perhaps ever before. Digital media here serve, for example, as architecture for collaborative platforms that structure pre-existing environmental knowledge. The past decade has seen a spread of such sharing platforms, some more successful than others, in their aim to function outside or independently of the capitalist economy. Regardless of their use merely as networks for social connectivity and interaction, or as collaborative networks that shape individual and group behaviour, it is becoming more evident that social media use (re)frames problems and have the capacity to also stimulate discussions. The expansion of networks based on the Global Positioning System (GPS) is also showing its impact on cities. On the one hand we see smart cities innovating with internet technology to enhance urban sustainability performance, for example, allowing neighbours to design low-carbon routes (Albino et al. 2015). But on the other hand, certain popular public platforms such as Amazon (online shopping), Uber (taxi service) and Airbnb (home lending) are becoming so powerful that they can change the identity and shape of a whole neighbourhood in a short time span. In the following sections we take a further look at the importance of place in terms of 'the local', 'the global' as well as 'the digital' in environmental communication of today.

The Local and Environmental Communication

'Place' is uniquely situated within many of the social sciences: geographical location is central to the shaping of identities, culture and belongingness. Each of us can think of a number of different places that define us and shape our identity—there are physical places such as those of living, work and worship, places of solitude, community places, as well as non-physical ones such as places where we can 'be ourselves'. As Williams and Brandt (2013, p. 44) write, 'these places are interwoven into the fabric of our social lives and indeed are integral to its possibility'. Place hence means both social and geographical place—for instance, 'out of place' (Cresswell 2004). Cresswell argues that place plays a key role in creating and continuing ideological beliefs.

Many of these places are determined by their naturalness, by the quality of their connection or disconnection with natural environments. For many tribal or Indigenous people this is especially true: all places, of living, work, worship and so on, are really one 'place' (hills, forests) that defines all other places and spaces. Furthermore, these places are key to shaping the way tribal or Indigenous peoples see themselves and their 'place' in the world. Places are thus defined by natural, social, cultural and personal processes. Certainly, environments as places and their quality impact identity, sense of self and how we communicate about the natural world. Environmental conflicts therefore are deeply tied to place. In 2012, Indigenous peoples' protests in eastern India against a large UK-based mining corporation were tied to the fear of destruction of lands considered sacred; placeness, that is, the quality of living and occupying a place, was central to this struggle (see Mishra 2013). However, urban environments in many parts of the world may change routinely so much that as Corbett (2006) suggests, familiarity with a place may also lead to an inability to see one's own place (and environment) as degraded.

Communicating about the environment has much to do with place and placeness. It is place that determines the nature of environmental communication, and it is environmental communication that determines the way a given place is understood. Environmental communication 'takes place in a geographic context' and people's thoughts of the environment are connected to who they are and 'where they exist' (Cantrill 2008, p. 179). While action on environmental issues is 'affixed to place and to the political context of the nation', the internet has allowed activists to 'locate their action within much broader movements' (Hansen 2010). This marks a fluidity in the understanding of the 'local', as environmental impacts are hardly restricted to a given place: impacts and repercussions of environmental action and inaction go beyond the local.

The same can be said for processes such as globalization, where the assumption of a bounded place as 'home' is difficult; what 'home' is itself and its possibilities are reconsidered, as it is 'networked to the globe' through telephones, televisions, the internet, air travel and so on, and is 'subject to the climatic, environmental and political affections of actions taken elsewhere' (Edwards and Usher 2008). Taylor (2010) argues that our connections to place 'might appear to have lessened in importance' in the twenty-first century with increased mobility. Increased mobility, therefore, 'has weakened the kind of traditional ties of place invoked in references to a native land, a birthplace, a home town or region, or a local community' (Taylor 2010, p. 2). Massey (1997) has argued that place should be seen as open and interconnected flows.

In this introductory chapter we look at place as one critical and complex dimension of environmental communication, with meanings beyond geography and interconnectivity. Far from attempting to define or delimit the concept, we observe it as a defining feature of environmental communication. Just as Meyrowitz (1986) analysed how new media affect the connection between physical place and social place, here we deliberately invoke 'the local' as well as 'the global' and 'the digital' when we refer to place. There are elements of what it means to invoke 'the local' that seem salient across time while the digital, as this volume attests, has also changed how local is understood and experienced today. Even anecdotally, thinking about the local and what this has meant in the past and what it means today points to shifts in how we create, receive and distribute communication about environmental issues. As Lester (2019, p. 4) observes, connections to our local places have transformed, as the pervasiveness in our everyday lives of communication and media brings global industries, events and issues to our homes, our communities and our society.

The caveat is that one's local place still matters to people and communities, evident every day in the deluge of news and media that tell stories of fierce, committed and passionate defence of where people live. And that this 'local place' persists alongside the change heralded by powerful global structures and their quotidian influence. As human communities, living symbiotically with nature, where we reside still matters, in terms of how we explain ourselves and place within the physical world. Reflections on 'place' have spanned diverse disciplines but especially cultural geography with the work of Carl Sauer and his Berkeley School. This work is well known for problematizing relations between human cultures, landscape and nature (see Johnson et al. 2013). Cultural geography emphasized 'the power of the geographical imagination in the process of social formation' (Cresswell 2010, p. 171). This is familiar terrain for environmental communication where research premised on investigations of (inter alia) 'place' and/or the social construction of environments or 'nature' in part defines the field (Cox and Depoe 2015; Hansen and Cox 2015).

Mapping and documenting these relations to place extends scholars beyond a 'way of seeing' to capture place as a 'way of experiencing' (Wattchow 2013, p. 90). In examining our local places we become the 'empathetic insider' (Relph 1976) to illustrate the rich dimensions of presence and how this can differ, bringing both stability and flux to definitions of how we understand where we are and who are—in our environments. The neutral bystander, observing from a distance, is supplanted by a richer, subjective immersion in our places. In 1974, Tuan coined the term 'topophilia' as 'the affective bond between people and place or setting' that he described as a 'diffuse concept' yet 'vivid and concrete as a personal experience' (Tuan 1974, p. 4). In a similar vein, Raymond Williams coined the phrase 'structure of feeling' which he defined as 'the felt sense of the quality of life at a particular place and time' (Williams 1975, p. 47). William's structure of feeling 'represents an attempt to integrate the notion of experience into social and cultural analysis' (Harding and Pribram 2004, p. 868). Other attempts to articulate our relationships to places include the vision of cities as places that are experienced and not merely as physical places. Sassen (2013) approaches cities as complex systems and recognizes their capacity to contest through their systemic dynamic. She claims that cities 'have speech, albeit informal and mostly unrecognized as such' (Sassen 2013, p. 210). Digital platforms are providing channels for these speeches in various forms. A number of chapters in this volume explore some of these experiences, where social networks in urban and more 'natural' contexts have developed ways of communicating their environment, of raising local protest voices or of capacity-building, to name some examples.

Communication scholars share the idea that 'place' has significance for understanding environments and the way environmental issues and conflicts come to be known. Evident in environmental communication research undertaken throughout the world, this truism is repeated and is echoed again in this volume. Communicating about environment entails an engagement with, and understanding of, place, placeness, and as often, 'placelessness' (see Relph 1976). It is place—in the sense of the complex relations between culture/s and nature/s—that is expressed in environmental communication, and it is environmental communication that represents the way a given place is understood. Cox's (2006) essay is instructive here, whereby environmental communication is defined as 'pragmatic' in the sense of education, awareness, leading to eventual solutions to environmental issues; and 'constitutive', as constructing representations of nature that enable us to identify and define problems, depending on our perspective. Drawing on these ideas, Foxwell-Norton (2018, p. 39) suggests we consider the constitutive dimensions of environmental communication as the a priori condition for our pragmatic responses to environmental issues. In other words, how we understand, frame and construct nature both enables and limits what we see and can possibly experience of our places. Our places 'become', in these shifting relations of the communication between culture and nature, which defines what we see and experience and in turn the range of possible responses and solutions to environmental issues.

The creation of digital networks and the capacity to communicate our local places has transformed the ways in which places are known and understood, adding a global dimension to the way we experience the 'local'. There is a 'new visibility' (Thompson 2005) of the places where we reside as 'locals' and those far distant from our immediate experiences. Thompson (2005, p. 35) draws on the ideas of 'place' and 'local' to explain the development of digital media communication. He argues that this mediated visibility of common, and not, locales is the way in which aspects of social and political life are brought to public prominence and, more significantly, the principal means 'by which social and political struggles are articulated and carried out'. Evidence of this transformation in how we understand and experience our local place—and through which we communicate ourselves and environment, environmental issues and conflict—is ubiquitous for the environmental communication scholar.

Contemporary environmental conflict and protest epitomizes how the experience of place is transformed by digital networks and yet still remains deeply tied to place. The enthusiasm for the potential of digitalization is tempered by the stark acknowledgement that inequalities in power, wealth and influence persist, regardless of the size, speed, distance and functionality of digital capabilities. We take a deeper look at these transformations in the following section.

The Digital and Environmental Communication

Environmental communication, like any other research field in media and communication, has evolved in the last decade or so at the same time as major structural transformations in the media landscape have been taking place, both globally and locally. Increased ownership concentration of media businesses is one important visible trend (Facht and Ohlsson 2019; Sjøvaag and Ohlsson 2019; Vizcarrondo 2013). Major changes in the content and form of the journalistic profession itself are yet another development, where buzzwords like 'multitasking' and 'convergence' have been used to describe new multimedia and multimodal competences required by journalists today (Larrondo Ureta and Peña Fernández 2018; Nygren 2008; Wiik and Nygren 2016). At the same time the number of specialized reporters in many newsrooms, like science and environmental journalists, is decreasing (Murcott and Williams 2013, p. 152).

Digitalization processes of different kinds have also been taking place both within and outside traditional media houses in this 'age of datafication', as Livingstone (2019) labels our time. The growth of studies on data journalism is only one type of area which reflects this development (Appelgren 2018; Hermida and Young 2019). Programming skills; increased use of statistics (Big Data) when reporting; new multimodal visualization techniques; and increased interactivity with the audience, who co-create their own stories (through the sources they choose), are all features of this new computational journalistic practice. Computational journalism that makes use of digital cartographies has proven to be of particular importance for environmental communication since, as Salovaara argues (2016, p. 827), these maps are good for visualizing 'environmental damage, human-made disaster, and endangered species' given that they 'include notions of species, space and territory that mark the move towards a new political ecology'. Digital maps are further understood as hybrid assemblages of mutually constituting environments that embrace the dependencies of nature and society, of human and non-human actors, hence their relevance for environmental communication researchers and practitioners.

The printed press, which up until now has been the main focus for environmental communication scholars (Comfort and Park 2018, cf. Schäfer 2012), is diminishing rapidly in many (Western) countries. The public, especially younger people, receive their news online to a much greater extent (Painter et al. 2018). And they do not only visit traditional national or regional/local media institutions' own webpages when doing so. Instead, newer 'digital-only' media players like *Huffington Post*, *BuzzFeed* and *Vice* have become popular news providers for younger audiences. Contemporary public(s) encounter news more and more often on social media platforms, such as Facebook and Twitter. These global platforms in turn are owned by gigantic, multinational tech firms, with few if any aspirations regarding the quality of the content of the published material (Campos-Freire et al. 2018; Nechushtai 2018). What this development means in terms of possibilities and hindrance for environmental communication has yet to be examined in detail. But there are definitely worrying signs when it comes to democratic implications of these structural changes, since they have an impact on political decision-making processes of importance for the environment. Indeed, local and regional media have been shown to provide (inter)national outlets in many countries with relevant information about places and processes that might be of more general national interest (Meadows et al. 2009). These communities and environments would remain invisible without such journalistic work and place-bound media coverage. Local news reporters have also proven to be most qualified when it comes to narrating more complex stories beyond the first-day take of an issue (Egan Sjölander 2016). And the literature has shown how community media are of particular relevance when it concerns environmental problems (Foxwell-Norton 2018; Howarth and Anderson 2019).

During the work with this edited volume we have tried to advance our understanding of what impact the growth of digital media use has had on environmental communication both locally and globally. Initially we raised the question whether it would make sense to talk about a 'digital turn' in our field during this third wave and we have also been interested in knowing if the digital means of communication seem to have altered our conceptual understanding and theoretical approaches to mediated environmental communication? When surveying the field and examining the literature from the last decade or so, one can conclude that such a turn has not happened, at least not yet, although there are enough signs of an inflection point. Already in 2015, climate change communication scholars like Koteyko et al. (2015, p. 149) pointed out the profound implications of the internet, and argued how it had given shape to a new 'communicative landscape'. They also pointed out new areas of study for climate change researchers interested in public engagement, highlighting the unprecedented possibilities of Web 2.0 and social media in particular.

Social media is also the most popular 'digital' subject among environmental communication studies. The majority focus on one medium, such as (web)blogs (Joosse and Brydges 2018; Poberezhskaya 2018), Facebook (Bloomfield and Tillery 2019; Katz-Kimchi and Manosevitch 2015; Olausson 2018), YouTube (Park 2018; Shapiro and Park 2018) or Twitter (Hutchins 2016; Jang and Hart 2015). Arlt et al.'s (2018) work about German citizens' online climate engagement constitutes an exception since it includes different kinds of social media. In Comfort and Park's (2018, p. 8) recent survey of the environmental communication field only 2 out of 50 top keywords are associated with the digital, namely, 'social media' and 'internet'. Neither of these are in the top tier either. Still we can observe that so-called passive crowdsourcing of social media for environmental research is indeed on the rise. This has been illustrated by Ghermandi and Sinclairs (2019, pp. 37–39) in their mapping of this emerging research field, by using social media data that have been generated for other purposes. These authors also claim that 'this new data source offers unprecedented opportunities to extend the scope, scale and depth of research, especially insofar as the interactions between humans and the environment are concerned' (Ibid, p. 43). In 2011 only one study was published using this kind of passive crowdsourced data, while in 2017 the number had increased to 61.

When reviewing environmental communication literature about 'the digital', we come across quite a few studies that present optimistic views like Lyon and Montgomery's (2013, p. 747) perspective that 'social media will reduce the incidence of corporate greenwash', or the launch of Bengston et al.'s (2009) new method for so-called rapid issue tracking of opinions about environmental issues as a service to decision-makers. Some cases also present progressive or at least promising results regarding the impact of digital communication like Hopke's (2016) analysis of what she calls 'translocal "broker" organizations' and their activism online against fracking, which connects very localized concerns about environmental risks with more abstract global ones. Arlt's research team (2018) about boosting climate change engagement from social media use is another such example, as is Collins and Nerlich's (2015) examination from a deliberative democracy perspective of user comments in climate change debates online using corpus linguistics (see also Hutchins 2016).

However, we also come across a range of studies that are more about unfulfilled promises regarding, for example, improved interactivity and dialogue online (cf. Abitbol et al. 2019; Poberezhskaya 2018). Comfort and Hester (2019, p. 281) likewise conclude that previous research shows that NGOs have often not been engaging in 'the dialogic potential of the platform' when using social media. Lee et al. (2018) come to a similar conclusion regarding the United States National Oceanic and Atmospheric Administration's (NOAA) interaction with publics in social media. In the same manner Adams and Gynnild (2013) argue that place sensitivity needs to be acknowledged in environmental messages in online media to a much greater extent.

The significance of place, as a product of the intertwining of the local and the global, in environmental discourses is also highlighted in Autry and Kelly's (2012) work on online public discourse about new investments in nuclear power. The pivotal connections between the global and the local have very much been built into the digital platform InfoAmazonia, an international environmental journalism network, which was launched in 2012 in order to improve the public's perception and support for a more sustainable development in the Amazon delta. Apart from sharing news coverage and reports from this region for free, the platform also 'provides open access to geo-referenced data on environmental changes within the region, in addition to tools to use and visualize these data' (Salovaara 2016, p. 830). In this and similar projects we can see that crowdsourced and collective data sharing enabled by digital means of communication can be of significant importance for environmental communication. Abstract risks such as global warming can, for example, be linked to concrete practices of deforestation and the removal of Indigenous peoples living in the Amazonas. The possibility to construct such large-scale views built from the tracking of locally mined data can of course also serve other purposes. In conjunction with all this 'Spreadable Media' (Jenkins et al. 2013) thriving in our networked cultures that prime the sharing of content, we have to keep in mind that there is always a backside to this coin, as Couldry and Mejias (2019) convincingly show in their book *The Costs of Connection: How Data Is Colonizing Human Life and Appropriating It for Capitalism*. The surveillance aspects enabled by the internet and Web 2.0 cannot be underestimated. Nor can the digital divides that emerge alongside this changing media landscape (Couldry et al. 2018a, b; Van Dijk and Hacker 2003).

User-generated content in the form of participatory journalism created by different publics online instead of professional journalists has often been promoted as an alternative to shrinking newsrooms (Holt and Karlsson 2015; Singer et al. 2011; Singer 2014). There is, however, especially when it comes to resource-dependent environmental reporting, very little empirical evidence that such voluntary practice can fill the hole that the previously paid, educated journalists leave behind (Egan Sjölander 2016). Another important aspect of the meaning of the digital when it comes to environmental communication concerns the material dimensions of this digital development, what Ulrich Beck (1992) would have classified as the unintended side effects of the production and use of modern technologies. All e-waste that we generate as media users is a prime example of the materiality of the digital. It also intertwines the global and the local in the most concrete and problematic way. In sum, rich Western countries depend on the Global South both as producers and as providers of computers, tablets and smartphones, at the same time as enormous amounts of electronic waste containing hazardous materials are exported to these poorer countries. Richard Maxwell and Toby Miller (2012) called for the urgent need of *Greening the media*, both in theory and in practice, based on a severe critique of our widespread ignorance about the environmental impact of the media industry. Unfortunately, contrary to what one then could expect when looking at our field today, there has not been that much work done in this area yet. Maxwell and Miller's forthcoming book *How Green Is Your Smartphone?* constitutes one important exception. Kääpä's (2018) book *Environmental Management of the Media: Policy, Industry, Practice* about the media industries in the UK and the Nordic countries is yet another, where the material environmental impact of these industries is critically reviewed.

To conclude, when trying to grasp the meaning of the digital in relation to environmental communication, we still see a strong need for different critical approaches that follow the wider consequences of a changing digital media landscape. And in the same manner as it makes good sense to relate the local and the global to each other analytically, one can see the value of applying similar approaches when it comes to understanding new versus old media. Because as Couldry and Van Dijk (2015) remind us of when analysing contemporary social media, traditional mass media have also always been 'social' institutions and provided infrastructures for the social lives of communities.

Where the Local Meets the Digital: Current Research and Future Challenges

In the previous pages, we have attempted to frame this volume and the field of environmental communication in a changing landscape, while reflecting on how to conceptualize developments regarding the (intertwining of the) local and the digital. But more importantly, it calls for the acknowledgement of a changing scenario for environmental communication that raises challenging, new questions for future research. The chapters that follow already give us some hints in that regard. They have been structured into three sections that provide direction as to where we are and where we might be heading.

In the first section, 'Digital Technologies, Place and Environmental Awareness', three chapters address how digital environmental monitoring is influencing environmental communication at the local level. We see digital technologies allying with place to make environmental problems visible, raise awareness and build capacity. Derek Moscato reflects on how spatial media such as drones with videography equipment can be used in combination with video streaming services to make *invisible* natural phenomena *visible*, and, for example, raise environmental awareness about remote endangered areas. Based on a discussion of a campaign by the British Columbia's Ancient Forest Alliance (AFA) on Vancouver Island, Canada, he shows how drone media activism can serve as a tool to level the playing field between activists and local communities on the one hand and state or corporate institutions on the other.

Both Kenneth Yang and Yowei Kang's chapter and Matteo Tarantino's make a similar argument about how digital technologies play an indispensable role for environmental NGOs in China to level the playing field with governmental institutions. Starting from an organizational capacity-building perspective, Yang and Kang show how social networking platforms and QR code applications on mobile phones have been used in conservation campaigns by environmental NGOs to mobilize resources and volunteers, enhance collaborations between one another and increase political participation among local stakeholders.

Similar to Moscato's, Tarantino's chapter also shows how mobile applications are used as a tool to make visible what is invisible and raise environmental awareness, in this case about the level of air pollution in Beijing. His analysis reveals how environmental NGOs redesign their mobile applications in a context in which air quality and pollution indexes serve as contested indicators between governmental institutions and environmental NGOs. The intertwining of place and digitalization manifests itself very clearly in this case in which the personal experience of a particular locality or space (in terms of air pollution) is the product of an interplay between material, symbolic and practice levels. The relationship between pollution and space is directly mediated by the design of digital applications.

The second section, 'Social Media, Environmental Protest and Community-Building', takes a look at the role of environmental communication regarding the emergence of local environmental activism and networking using social media. Anders Horsbøl's chapter analyses how local communities in Denmark protest the construction of wind turbines *with* and *in* Facebook groups. He argues that the meeting of the digital and the local should not be understood as the connection of two separate realms, but as the intertwining of two types of connectivity: he finds the Facebook groups functioning as a space where non-local events, issues and discourses are being domesticated and digitized simultaneously in complex ways. Likewise, René Jara and Claudio Broitman focus on how activists use social media such as Facebook fan pages to challenge legacy media's preference for official discourses by governmental and scientific representatives. They illustrate this by showing how activists, in the case of Chili's Chiloé Island's red tide phenomenon, succeeded in turning what was first perceived as a natural disaster into an environmental crisis related to local salmon farming and fishing industries.

This corresponds to the chapter by Tracy Mwaga Tinga, Patrick Murphy and Emmanuel Septime Sessou, which examines the adoption of digital applications in sub-Saharan Africa: it reveals how these applications are reshaping the ways in which citizens are able to respond to the challenges of environmental problems, for example, by opening up opportunities for more regionalized and localized community-building, knowledge sharing and action. This is illustrated by several case studies, such as the VUMA. EARTH call-to-action campaign website launched by Greenpeace Africa, the use of mobile phone apps by farming communities in Kenya and a highly networked communication effort behind the social media stewardship campaign Water 4 Cape Town.

The chapter by Camila Emboava Lopes and Annika Egan Sjölander focuses on how Indigenous media activists in Brazil use Facebook, Instagram and Twitter to construct their contemporary indigeneity, in their own terms, online. This (re)articulation of their identities in relation to ideas of land, nature and justice aims to challenge hegemonic stereotypes and dominant discourses that seek to undermine their legitimacy and power. The chapter shows how the digital context and its specific means of communication affect the dynamics between the local, national and global level in the discourse, making power relations more complex compared to before the wide introduction and use of digital and social media.

In the final section, 'Digital Communication, Environmental Discourse and Local Experiences', the chapters introduce connections between the digital and its impact on communication and sustainable living. Alana Mann's chapter reflects on how digital technologies have the potential to redefine the relationships and connections between citizens and (city) governments towards more inclusive and participatory food governance. She sees many opportunities for digital food activism in a context in which digital platforms give consumers access to information on food, facilitating the emergence of more resilient and inclusive food networks, and counter the unsustainable food practices of Big Food in the process.

Chamee Yang's chapter on the other hand challenges the role of digital technologies in the local experience of a smart city. Drawing from the case study of the South Korean city of Songdo, she shows how digital technologies, and datafication in particular, can be used to impose standardized universal solutions that serve business interests, first and foremost, to what are really problems of social relations in an urban environment. In doing so, she problematizes the interplay between the historical evolution of sustainable development as a new growth paradigm and the new urban governance model of smart cities afforded by the new regime of digital media.

Franzisca Weder takes a different approach by focusing on the potential for local environmental journalism to transform into a form of constructive/solution journalism. For instance, when journalists succeed in engaging with Facebook to domesticate environmental issues and reduce their complexity, subsequently creating a (hyper)local digital communication space that enables community engagement. She illustrates this by discussing ten different case studies related to sustainable water management and the response of local communities to water scarcity in Europe, Australia, South Africa and the US.

In their Australian case study of home renovations and energy efficiency, Aneta Podkalicka and Deb Anderson call for a situated integration of media in everyday life, and more specifically, for a focus on how environmental communication and media practices (ranging from instructional DIY YouTube videos, over Facebook groups, to international reality TV programmes) are experienced in their lived, local and everyday cultural context. Regardless of the medium, lived experience in terms of local knowledge and social talk is found to be generated and valued as a useful resource to guide people through the renovation process. Starting from a culturalist perspective, their chapter demonstrates how local places and digital spaces are interconnected and serve as resources for navigating the abundance and fragmentation of available information.

One common feature of the experiences analysed in these chapters is that they have quite an optimistic view of the intertwining of the local and the digital in environmental communication. Nevertheless, we argue that there is still a need to dig deeper into the 'darker' side of digital technologies and the challenges posed to environmental communication, associated with their use and impact, on a local-global scale. The first section looks at instrumental opportunities for the local and the appropriation of the digital by community actors. At the same time, we identify a need for research that clarifies how environmental communication, for example, can contribute to differentiate monitoring for environmental awareness from surveillance to environmental injustice. The second section explores digital networking as a new strategy for environmental activism at the community level. But more research is needed regarding the identification and understanding of the roles, practices and power of both local and global actors. Who is really leading environmental digital activism? Are these new forms of protest evidence of a transformational use of environmental communication? The chapters in this section acknowledge the salience of place and proximity as the markers of environmental discourse in social media. However, we will have to further understand environmental communication under the digital logic of claims-making and how it uses place as a vindicator of environmental rights and justice. And the third section offers insights into how environmental communication is challenged when the local and the digital intersect as modes of social and political participation. From these case studies we also are reminded about the importance of context when understanding and analysing environmental communication.

The entire volume bears witness to how the local and the digital connect environmental communication research and practice with broader questions regarding democracy, governance and the political. These questions demand attention from environmental communication scholars. It is clear also that digitalization processes redefine the role and power of communities, places, cities and regions in relation to government/s and transnational corporations. The environmental crisis we face today challenges scholars and the field to reflect further upon how the digital can be used to enhance sustainability in all its forms. And this instead of being technologies operating in the interests of powers that exploit people, the environment, our common goods and resources.

References

Abitbol, A., Meeks, J., & Cummins, R. G. (2019). Does oil and goodwill mix?: Examining the oil and gas industry's impact on stakeholder engagement on Facebook. *Environmental Communication, 13*(2), 192–208.

Adams, P. C., & Gynnild, A. (2013). Environmental messages in online media: The role of place. *Environmental Communication: A Journal of Nature and Culture, 7*(1), 113–130.

Albino, V., Berardi, U., & Dangelico, R. M. (2015). Smart cities: Definitions, dimensions, performance, and initiatives. *Journal of Urban Technology, 22*(1), 3–21.

Anderson, A. (1997). *Media, culture and the environment*. London: UCL Press.

Anderson, A. (2014). *Media, environment and the network society*. London: Palgrave Macmillan.

Appelgren, E. (2018). An illusion of interactivity: The paternalistic side of data journalism. *Journalism Practice, 12*(3), 308–325.

Arlt, D., Hoppe, I., Schmitt, J. B., De Silva-Schmidt, F., & Brüggemann, M. (2018). Climate engagement in a digital age: Exploring the drivers of participation in climate discourse online in the context of COP21. *Environmental Communication, 12*(1), 84–98.

Autry, M. K., & Kelly, A. R. (2012). Merging Duke Energy and Progress Energy: Online public discourse, post-Fukushima reactions, and the absence of environmental communication. *Environmental Communication: A Journal of Nature and Culture, 6*(2), 278–284.

Beck, U. (1992). *Risk society: Towards a new modernity*. London: Sage.

Bell, A. (1994). Media (mis)communication on the science of climate change. *Public Understanding of Science, 3*, 259–275.

Bengston, D. N., Fan, D. P., Reed, P., & Goldhor-Wilcock, A. (2009). Rapid issue tracking: A method for taking the pulse of the public discussion of environmental policy. *Environmental Communication, 3*(3), 367–385.

Bloomfield, E. F., & Tillery, D. (2019). The circulation of climate change denial online: Rhetorical and networking strategies on Facebook. *Environmental Communication, 13*(1), 23–34.

Burke, K. (1966). *Language as symbolic action: Essays on life, literature, and method*. Berkeley: University of California Press.

Campos-Freire, F., de Aguilera-Moyano, M., & Rodríguez-Castro, M. (2018). The impact of global platforms on media competition and on the results of European communication companies. *Communications and Society, 31*(3), 223–238.

Cantrill, J. G. (2008). A sense of self-in-place for adaptive management, capacity building and public participation. In S. L. Senecah (Ed.), *The environmental communication yearbook* (Vol. 1, pp. 179–200). New York: Routledge.

Carvalho, A. (2007). Ideological cultures and media discourses on scientific knowledge: Re-reading news on climate change. *Public Understanding of Science, 16*(2), 223–243.

Castells, M. (2009). *Communication power*. Oxford: Oxford University Press.

Catton, W. R., & Dunlap, R. E. (1978). Environmental sociology: A new paradigm. *The American Sociologist, 13*, 15–47.

Chapman, G. (1997). *Environmentalism and the mass media: The North–South divide* (Global environmental change series). London/New York: Routledge.

Clark, J. D., & Themudo, N. S. (2006). Linking the web and the street: Internet-based "dotcauses" and the "anti-globalization" movement. *World Development, 34*(1), 50–74.

Collins, L., & Nerlich, B. (2015). Examining user comments for deliberative democracy: A corpus-driven analysis of the climate change debate online. *Environmental Communication, 9*(2), 189–207.

Comfort, S. E., & Hester, J. B. (2019). Three dimensions of social media messaging success by environmental NGOs. *Environmental Communication, 13*(3), 281–286.

Comfort, S. E., & Park, Y. E. (2018). On the field of environmental communication: A systematic review of the peer-reviewed literature. *Environmental Communication, 12*(7), 862–875.

Corbett, J. B. (2006). *Communicating nature: How we create and understand environmental messages*. Washington, DC: Island Press.

Couldry, N., & Mejias, U. A. (2019). *The costs of connection: How data is colonizing human life and appropriating it for capitalism*. Stanford: Stanford University Press.

Couldry, N., & Van Dijck, J. (2015). Researching social media as if the social mattered. *Social Media Society, 1*(2), 1–7.

Couldry, N., et al. (2018a). Media and communications. In International Panel on Social Progress (IPSP) (Ed.), *Rethinking society for the 21st century* (pp. 523–562). Cambridge: Cambridge University Press.

Couldry, N., et al. (2018b). Media, communication and the struggle for social progress. *Global Media and Communication, 14*(2), 173–191.

Cox, J. R. (2006). *Environmental communication and the public sphere*. Thousand Oaks: Sage.

Cox, R., & Depoe, S. (2015). Emergence and growth of the 'field' of environmental communication. In A. Hansen & R. Cox (Eds.), *The Routledge handbook of environment and communication* (pp. 13–25). Oxford: Routledge.

Cresswell, T. (2004). *Place: A short introduction*. Malden: Blackwell Publishing.

Cresswell, T. (2010). New cultural geography – An unfinished project? *Cultural Geographies, 17*(2), 169–174.

Deetz, S. (1995). *Transforming communication, transforming business: Building responsive and responsible workplaces* (The organizational experience in modern society). Cresskill: Hampton Press.

Delicath, J. W., Elsenbeer, M. A., & Depoe, S. P. (2004). *Communication and public participation in environmental decision making*. Albany: SUNY Press.

Díaz-Pont, J. (2000). *Sostenibilitat en una òptica global. Economia, empresa i societat*. Barcelona: Beta.

Díaz-Pont, J. (2017). Mapping the changes of CSR stakeholder perceptions in a context of economic depression in Spain: An approach from communication and governance. In L. M. Gómez et al. (Eds.), *Corporate social responsibility and corporate governance: Concepts, perspectives and emerging trends in Ibero-America* (pp. 89–110). Croydon: Emerald Publishing Limited.

Durant, J. R., Evans, G. A., & Thomas, G. P. (1989). The public understanding of science. *Nature, 340*(6228), 11–14.

Edwards, R., & Usher, R. (2008). *Globalisation and pedagogy: Space, place and identity* (2nd ed.). London: Routledge.

Egan Sjölander, A. (2016). Journalistik och komplexa beslut. In O. Westerlund (Ed.), *Människorna, medierna och marknaden: medieutredningens forskningsantologi om en demokrati i förändring* (pp. 351–375). Stockholm: Wolters Kluwer.

Facht, U., & Ohlsson, J. (2019). *Medie-Sverige 2019*. Göteborg: Nordicom.

Foxwell-Norton, K. (2018). *Environmental communication and critical coastal policy: Communities, culture and nature*. London: Routledge.

Gamson, W. A., & Modigliani, A. (1989). Media discourse and public opinion on nuclear power: A constructionist approach. *American Journal of Sociology, 95*(1), 1–37.

Ghermandi, A., & Sinclair, M. (2019). Passive crowdsourcing of social media in environmental research: A systematic map. *Global Environmental Change, 55*, 36–47.

Hajer, M. A., & Fischer, F. (1999). Living with nature: Environmental politics as cultural discourse. In *Oxford*. New York: Oxford University Press.

Hansen, A. (1993). *The mass media and environmental issues* (Studies in communication and society). Leicester: Leicester University Press.

Hansen, A. (2010). *Environment, media and communication*. Abingdon/New York: Routledge.

Hansen, A., & Cox, R. (2015). Introduction: Environment and communication. In A. Hansen & R. Cox (Eds.), *The Routledge handbook of environment and communication* (pp. 1–12). Oxford: Routledge.

Harding, J., & Pribram, E. D. (2004). Losing our cool? Following Williams and Grossberg on emotions. *Cultural Studies, 18*(6), 863–883.

Hermida, A., & Young, M. L. (2019). *Data journalism and the regeneration of news*. London: Routledge.

Holt, K., & Karlsson, M. (2015). "Random acts of journalism?": How citizen journalists tell the news in Sweden. *New Media & Society, 17*(11), 1795–1810.

Hopke, J. E. (2016). Translocal anti-fracking activism: An exploration of network structure and tie content. *Environmental Communication, 10*(3), 380–394.

Howarth, C., & Anderson, A. (2019). Increasing local salience of climate change: The un-tapped impact of the media-science interface. *Environmental Communication, 13*(6), 713–722. https://doi.org/10.1080/17524032.2019.1611615.

Hutchins, B. (2016). The many modalities of social networking: The role of Twitter in greens politics. *Environmental Communication, 10*(1), 25–42.

Jang, S. M., & Hart, P. S. (2015). Polarized frames on "climate change" and "global warming" across countries and states: Evidence from Twitter big data. *Global Environmental Change, 32*, 11–17.

Jenkins, H., Ford, S., & Green, J. (2013). *Spreadable media: Creating value and meaning in a networked culture*. New York: New York University Press.

Jessop, B. (2016). Territory, politics, governance and multispatial metagovernance. *Territory, Politics, Governance, 4*(1), 8–32.

Johnson, N. C., Schein, R. H., & Winders, J. (2013). Introduction. In N. C. Johnson, R. H. Schein, & J. Winders (Eds.), *The Wiley-Blackwell companion to cultural geography* (pp. 26–39). New York: Wiley.

Joosse, S., & Brydges, T. (2018). Blogging for sustainability: The intermediary role of personal green blogs in promoting sustainability. *Environmental Communication, 12*(5), 686–700.

Kääpä, P. (2018). *Environmental management of the media: Policy, industry, practice*. London: Routledge.

Kamieniecki, S. (1993). *Environmental politics in the international arena: Movements, parties, organizations, and policy*. Albany: State University of New York Press.

Katz-Kimchi, M., & Manosevitch, I. (2015). Mobilizing Facebook users against Facebook's energy policy: The case of Greenpeace unfriend coal campaign. *Environmental Communication, 9*(2), 248–267.

Kilbourne, W. E. (2004). Sustainable communication and the dominant social paradigm: Can they be integrated? *Marketing Theory, 4*(3), 187–208.

Koteyko, N., Nerlich, B., & Hellsten, I. (2015). Climate change communication and the internet: Challenges and opportunities for research. *Environmental Communication, 9*(2), 149–152.

Kunst, M., & Witlox, N. (1993). Communication and the environment. *Communication Research Trends, 13*(1), 1–31.

Larrondo Ureta, A., & Peña Fernández, S. (2018). Keeping pace with journalism training in the age of social media and convergence: How worthwhile is it to teach online skills? *Journalism, 19*(6), 877–891.

Lee, N. M., Van Dyke, M. S., & Cummins, R. G. (2018). A missed opportunity?: NOAA's use of social media to communicate climate science. *Environmental Communication, 12*(2), 274–283.

Lester, L. (2010). *Media and environment: Conflict, politics and the news.* Cambridge: Polity Press.

Lester, L. (2019). *Global trade and mediatised environmental protest: The view from here.* Cham: Palgrave Macmillan.

Lischinsky, A., & Egan Sjölander, A. (2014). Talking green in the public sphere: Press releases, corporate voices and the environment. *Nordicom Review, 35*(Special Issue), 125–139.

Livingstone, S. (2019). Audiences in an age of datafication: Critical questions for media research. *Television & New Media, 20*(2), 170–183.

Lyon, T. P., & Montgomery, A. W. (2013). Tweetjacked: The impact of social media on corporate greenwash. *Journal of Business Ethics, 118*(4), 747–757.

Maeseele, P. (2015). Beyond the post-political zeitgeist. In H. Hansen & R. Cox (Eds.), *The Routledge handbook of environment and communication* (pp. 429–443). New York: Routledge.

Martin, C. J. (2016). The sharing economy: A pathway to sustainability or a nightmarish form of neoliberal capitalism? *Ecological Economics, 121*, 149–159.

Massey, D. (1997). A global sense of place. In T. Barnes & D. Gregory (Eds.), *Reading human geography.* London: Arnold.

Maxwell, R., & Miller, T. (2012). *Greening the media.* Oxford: Oxford University Press.

Mazur, A. (1984). The journalists and technology: Reporting about love canal and three Mile Island. *Minerva, 22*, 45–66.

Mazur, A., & Lee, J. (1993). Sounding the global alarm: Environmental-qüestions in the United-States national news. *Social Studies of Science, 23*(4), 681–720.

Meadows, M., Forde, S., Ewart, J., & Foxwell, K. (2009). Making good sense: Transformative processes in community journalism. *Journalism, 10*(2), 155–170.

Meyrowitz, J. (1986). *No sense of place: The impact of electronic media on social behavior*. New York: Oxford University Press.

Mishra, M. (2013). Communicating the true ecological cost of development: Addressing development and environment in Orissa, India. In J. Servaes (Ed.), *Sustainable development and green communication* (pp. 147–169). London: Palgrave Macmillan.

Muir, S. A., & Veenendall, T. L. (1996). *Earthtalk: Communication empowerment for environmental action* (Praeger series in political communication). Westport: Praeger.

Murcott, T. H., & Williams, A. (2013). The challenges for science journalism in the UK. *Progress in Physical Geography, 37*(2), 152–160.

Nechushtai, E. (2018). Could digital platforms capture the media through infrastructure? *Journalism, 19*(8), 1043–1058.

Newell, P., et al. (2015). Governance traps in climate change politics: Re-framing the debate in terms of responsibilities and rights. *Wiley Interdisciplinary Reviews: Climate Change, 6*(6), 535–540.

Nisbet, E. C., Cooper, K. E., & Ellithorpe, M. (2014). Ignorance or bias? Evaluating the ideological and informational drivers of communication gaps about climate change. *Public Understanding of Science, 24*(3), 285–301.

Nygren, G. (2008). *Yrke på glid – om journalistrollens deprofessionalisering* [Profession in transition – On the deprofessionalisation of journalism]. Stockholm: Institutet för Mediestudier.

Olausson, U. (2018). "Stop blaming the cows!": How livestock production is legitimized in everyday discourse on Facebook. *Environmental Communication, 12*(1), 28–43.

Painter, J., Kristiansen, S., & Schäfer, M. S. (2018). How 'digital-born' media cover climate change in comparison to legacy media: A case study of the COP 21 summit in Paris. *Global Environmental Change, 48*, 1–10.

Park, H. W. (2018). YouTubers' networking activities during the 2016 South Korea earthquake. *Quality & Quantity, 52*(3), 1057–1068.

Poberezhskaya, M. (2018). Blogging about climate change in Russia: Activism, scepticism and conspiracies. *Environmental Communication, 12*(7), 942–955.

Relph, E. (1976). *Place and placeless-ness*. London: Pion Limited.

Salovaara, I. (2016). Participatory maps: Digital cartographies and the new ecology of journalism. *Digital Journalism, 4*(7), 827–837.

Sassen, S. (2013). Does the city have speech? *Public Culture, 25*(2 70), 209–221.

Schäfer, M. S. (2012). Online communication on climate change and climate politics: A literature review. *Wiley Interdisciplinary Reviews: Climate Change, 3*(6), 527–543.

Schultz, F., Castelló, I., & Morsing, M. (2013). The construction of corporate social responsibility in network societies: A communication view. *Journal of Business Ethics, 115*(4), 681–692.

Sen, S., Bhattacharya, C. B., & Korschun, D. (2006). The role of corporate social responsibility in strengthening multiple stakeholder relationships: A field experiment. *Journal of the Academy of Marketing Science, 34*(2), 158–166.

Shapiro, M. A., & Park, H. W. (2018). Climate change and YouTube: Deliberation potential in post-video discussions. *Environmental Communication, 12*(1), 115–131.

Singer, J. B. (2014). User-generated visibility: Secondary gatekeeping in a shared media space. *New Media & Society, 16*(1), 55–73.

Singer, J. B., et al. (2011). *Participatory journalism: Guarding open gates at online newspapers*. Chichester: Wiley.

Sjøvaag, H., & Ohlsson, J. (2019). Media ownership and journalism. In *Oxford research encyclopedia of communication*. Oxford: Oxford University Press.

Stevenson, H., & Dryzek, J. S. (2013). *Democratizing global climate governance*. Cambridge: Cambridge University Press.

Swyngedouw, E. (2017). Unlocking the mind-trap: Politicising urban theory and practice. *Urban Studies, 54*(1), 55.

Taylor, S. (2010). *Narratives of identity and place*. Hove: Routledge.

Thompson, J. B. (2005). The new visibility. *Theory Culture Society, 22*(6), 31–51.

Tuan, Y. F. (1974). *Topophilia: A study of environmental perception, attitudes, and values*. Englewood Cliffs: Prentice Hall.

United Nations, World Commission on the Environment and Development (WCED). (1992). *The Rio Declaration on Environment and Development*.

Van Dijk, J., & Hacker, K. (2003). The digital divide as a complex and dynamic phenomenon. *The Information Society, 19*(4), 315–326.

Vizcarrondo, T. (2013). Measuring concentration of media ownership: 1976–2009. *JMM: The International Journal on Media Management, 15*(3), 177–195.

Wattchow, B. (2013). Landscape and a sense of place: A creative tension. In P. Howard et al. (Eds.), *The Routledge companion to landscape studies* (pp. 87–96). New York: Routledge.

Wiik, J., & Nygren, G. (2016). Den journalistiska professionen i nätverksamhället. In O. Westerlund (Ed.), *Människorna, medierna och marknaden: medieutredningens forskningsantologi om en demokrati i förändring* (pp. 261–282). Stockholm: Wolters Kluwer.

Williams, R. (1975). *The long revolution*. Westport: Greenwood.

Williams, D. L., & Brandt, E. A. (2013). Sense of place, identity, and cultural continuity in an Arizona community. In P. N. Goggin (Ed.), *Environmental rhetoric and ecologies of place* (pp. 58–69). Abingdon/New York: Routledge.

Wood, D. J., & Logsdon, J. M. (2001). Theorizing business citizenship. In J. Andriof & M. McIntosh (Eds.), *Perspectives on corporate citizenship* (pp. 83–116). London: Greenleaf.

SECTION I

Digital Technologies, Place and Environmental Awareness

Aerial Activism in Old-Growth Rainforest: The Ancient Forest Alliance's Drone Videography as Spatial Communication

Derek Moscato

Introduction

"If a tree falls in the forest, does anybody hear?"—the eternal question is posed by musician and environmentalist Bruce Cockburn in the chorus of his 1989 song *If a Tree Falls*, which focuses on the deforestation of the Amazon and other threatened landscapes. Based on an older variation of the eco-philosophical adage featured in Mann and Twiss's book *Physics* (1910), it emphasizes the importance of bearing witness to ecological events, while turning attention to observed but also unobserved natural phenomena.

During the past decade, aerial technology has furthered humankind's ability to more fully experience the natural world, in turn creating an audience for contemporary environmental issues and debates. Unmanned

D. Moscato (✉)
Department of Journalism, Western Washington University, Bellingham, WA, USA
e-mail: Derek.Moscato@wwu.edu

J. Díaz-Pont et al. (eds.), *The Local and the Digital in Environmental Communication*, Global Transformations in Media and Communication Research - A Palgrave and IAMCR Series,
https://doi.org/10.1007/978-3-030-37330-6_2

aerial vehicles (UAVs), better known as drones, have previously enjoyed significant usage for military, security, and engineering purposes. Equipped with videography equipment such as GoPro cameras, they play a growing role in the environmental arena in activities such as locating wildlife, tracking natural disasters such as wildfires, and mapping remote geographies. The surveying of forests and other ecosystems by unmanned aerial vehicles is relatively low-cost and easy to learn, making it an increasingly practical and accessible tool for conservationists (Koh and Wich 2012). Environmental activists have been well-positioned to follow this lead. This is evidenced by examples such as the tracking of Japanese whaling ships (Zenko 2012; Simpson 2015) and turtle egg poaching (Moore 2015).

This study examines the 2015 campaign by British Columbia's Ancient Forest Alliance (AFA) to stop the heli-logging of old-growth forest on Vancouver Island's Central Walbran Valley in British Columbia, Canada. AFA's drone-captured video of pristine rainforest in a remote valley landscape represents a new approach to monitoring, documenting, and publicizing the ecological contestations and otherwise inaccessible terrain in this area. Drawing from the media theory of spatialization developed by political economist and communication scholar Harold Innis, this chapter explores the new interplay between technology, media, and ideology that the rise of drone media activism in the environmental sphere represents.

The Promise of Drone Media

Since their introduction, drones have provided government agencies and related organizations with a key technology for military, transport, surveillance, and mapping purposes. Such a technology has been used in military missions for reconnaissance, surveillance, and target acquisition (Pereira et al. 2009). While their early development was primarily driven by military utility and needs (Franke 2015b), in recent years they have become an important tool for both established media organizations and citizen journalists—and their utility in media production and meaning-making has been recognized by environmental activists. Even as the technology now features in better-resourced newsrooms because of its usefulness in producing visual journalism, Howley (2018) highlights how the drone is also rendered as the subject of novel media discourses: "Like the anxiety-laden news stories of near misses between drones and commercial airplanes, videos depicting animal aggression toward drones, such as the clip of a kangaroo punching a drone out of the sky, frequently garner global media attention" (p. 19). Drones, then, help to produce media discourses but are also the focal point of them.

There are several practical dimensions of drone use that make them a useful news gathering tool for media: their portable nature, their relatively low cost, and their safety in terms of gathering footage in dangerous conflict or weather events (Tremayne and Clark 2014). Their deployment in civil missions extends to aerial gravimetry, surveillance of coastal and maritime traffic, fishing surveillance, and detection of natural or coastal hazards (Pereira et al. 2009). The video drone, therefore, exists at the intersection of geography, technology, and media—providing visual aerial documentation of ecological phenomena.

Drawing from diffusion theory, Gynnild (2014) likens the innovation of drones for journalistic purposes to a "disruptive innovation," both because it has changed existing conceptions of visual journalism and because its emergence has given way to new markets and value networks within journalistic practice. Another perspective situates the role of UAVs in visual journalism as transformative, but places this development alongside a broader array of technological innovations for the newsroom, including smartphones, commercial satellite imagery, and cameras that are smaller and cheaper (Corcoran 2014).

There are limits to the usage of UAVs for conducting journalistic research or reporting, however. One of the most significant is the assurance of compliance for manned aircraft to strict air safety requirements—air worthiness requirements—which potentially precludes the presence of drones in non-segregated airspaces (Pereira et al. 2009). The utilization of drones in civil missions can therefore be mitigated by technical but also regulatory factors. In most countries and jurisdictions, the usage of UAVs is governed by a framework of jurisdictional-specific regulations. In the United States, for example, their usage is governed by the Federal Aviation Administration (FAA). As a result, national or regional regulatory environments greatly influence how, when, and where drones can be operated by journalists, media outlets, and private citizens (Holton et al. 2015). A more restrictive regulatory environment for UAVs—particularly for journalistic videography purposes—can create a tension with existing free speech laws such as the First Amendment in the United States (Love et al. 2015).

Another challenge for journalists is to incorporate such technology within a framework of ethical and safety considerations. Issues of privacy and surveillance remain paramount. This puts the onus on newsrooms to articulate ethical guidelines and transparency with the public for their usage, as well as instilling in journalists the importance of ethical reasoning within their reporting (Culver 2014). These emergent questions related to surveillance journalism are important within environmental communication, because they situate new media as conduits to engaging with ecological conditions in new and unprecedented ways.

From Surveillance Journalism to Environmental Activism

Aerial perspectives of the environment, such as those gathered by sophisticated satellite technology, offer scientists access to new information and data about land and marine ecosystems. Satellite imagery, for example, enables observers to detect changes to the environment, including changes in wildlife populations, undisclosed mining activity, and deforestation (Conniff 2017). While geospatial tools such as Google Earth may offer limited value to how citizens experience their own geography, they can provide distant viewers with the opportunity to witness changes to the earth as a result of industrial infrastructure such as energy projects, thus providing an important informational tool for conservation campaigns (Phadke 2010).

Discovery of the unknown emerges as a catalyst for the production and mediation of aerial imagery. A study of news stories featuring drone footage or reporting from 2010 to 2012 showed that many of the cases involved either secretive government actions or anti-government activism (Tremayne and Clark 2014). These examples included a secret detention centre in Australia and police-protest clashes in Poland, Russia, and the United States. Drones have similarly provided visual evidence of violence, corruption, or other forms of law-breaking. One example stems from the case of a Dallas man who flew his drone over a large industrial area, which included a meat-packing plant, only to later discover that the images he captured depicted pig blood being dumped into a river (Tremayne and Clark 2014). The photographer, described as a "UAV enthusiast," contacted local authorities, which resulted in a criminal investigation by the Environmental Protection Agency, the Texas Commission on Environmental Quality, and Texas Parks and Wildlife (Sidik 2012). Such a case demonstrates potential conflict between powerful interests and watchdog organizations: "It is not difficult to imagine drones being put to use by activist groups such as Greenpeace and others to capture images of things people with money and power may or may not want recorded" (Tremayne and Clark 2014, p. 239). It also signals the emergent theme of anti-authoritarianism within video drone operation, and the uptake of UAV media beyond newsrooms or other traditional media organizations.

In 2011, Sea Shepherd Conservation Society outfitted its two anti-whaling ships with drones in order to both track down and continually monitor Japanese whaling vessels (Heimbuch 2011). Yet Sea Shepherd's usage is an outcome of the popularity of drones in environmental and conservation circles. Sandbrook (2015) maintains that UAVs offer a more accurate and affordable technology for biodiversity conservation professionals—despite social impacts related to privacy, safety, and psychological well-being. In order to ensure that such negative impacts do not undermine the ecological benefits of drone technology, an integration of ethics and self-regulation for the operation of drones is therefore called for (Sandbrook 2015). Similarly, the growing uptake of UAVs by a multitude of organizations, including activists and media, has fostered calls for a framework to guide their usage. Choi-Fitzpatrick (2014) calls for drone media that fulfils a public good, a policy that simultaneously respects privacy and data while abiding by a "do no harm" principle.

Underpinning the elevated role of drones in contemporary mediation of the environment is the technology's ability to capture aerial video. As previous examples suggest, drone videography can be particularly adept at gathering evidence and information about changing material conditions and on-the-ground realities—in great part because the technology can often go where people, or other devices, cannot. The advent of digital and networked technologies has given activists the ability to bring ecological conflicts into open view and to make visible through mediation what was previously invisible (Lester and Hutchins 2012). At the same time, the images generated by drone videography can serve a rhetorical purpose. "Image events" are described by DeLuca (2005) as powerful, visualized mediations of the environment that allow activists to level the playing field with opposing institutions and governments. Image-based persuasive appeals can also elicit emotional responses from audiences because they connect visuals to existing cultural values. Gilliam Jr. and Bales (2002) call for rhetorical visual appeals—including photos, illustrations, maps, cartoons, charts, and other graphical representations—to be part of a larger strategic framing taxonomy for social and environmental advocates. Such images are rhetorically more effective than written text because they produce more immediate and visceral responses (Hill 2012). This view is echoed by Blair (2012) in emphasizing the power and possibility of visual arguments within societal debates: "It adds drama and force of a much greater order. Beyond that it can use such devices as references to cultural icons and other kinds of symbolism, dramatization and narrative to make a powerfully compelling case for its conclusion. The visual has an immediacy, a verisimilitude, and a concreteness that help influence acceptance" (p. 59). An example comes from Charles (2016) in describing the first close-up pictures of the planet Pluto by NASA's New Horizons spacecraft in July of 2015. Some media coverage in the British press pointed to a heart-shaped area depicted in the imagery—giving way to emotional, personalized narratives invoking words like "love" and "adore" (Charles 2016). While such a news narrativization by the traditional media added to the significance of the images, the event was also a milestone moment for social and digital media. Charles (2016) points to several parties, including NASA and US President Barack Obama, contributing to the virality of the persuasive spatial images by disseminating them via Twitter.

Spatial media have thrived in recent years as a result of the proliferation of new media technologies such as UAVs, as well as digital networks and devices. Through pictures, statistics, historical documents, reviews, and routes, they complement geographic spaces and are helping to transform both the production of space and the nature of spatiality (Kitchin et al. 2017). Increasingly, through the production of georeferenced data or metadata, such media help to make sense of geographic spaces and the interactions occurring within them. As a result, these same media help to heighten the symbolic meaning of spaces. They also help amalgamate the material and virtual worlds, creating hybrid spaces offering "dynamic and mutually constitutive" entanglements between the physical and digital worlds (Zook and Graham 2007). This parallels the discursive-material knot described by Carpentier (2017)—an "intense and inseparable" entanglement—in understanding community media participation. Howley (2018) similarly emphasizes the relationship between discursive and material approaches in understanding the role of drones in the popular imagination, given their real and imaginary roles in domestic airspace and popular imagination.

A case from Indonesia illustrates the power embedded in such technologies. Radjawali and Pye (2017) describe ongoing political interventions against palm oil and mining companies in the ecological transformation of the Kapuas River, the country's longest river, thanks to local communities' usage of drone-enabled counter-mapping. That UAVs can play such a significant role in bolstering communities' land rights also raises questions about their larger politicization and the inevitable disputes over their ultimate control.

Spatialization and the Mediation of the Hyperlocal

Long-standing conceptualizations of space and time offer key insights for environmental media scholarship. The term *spatialization* as described by Lefebvre and Enders (1976) highlighted the role of space and time in social life. A related understanding had been connected to media systems and mass communication industries decades earlier by Canadian historian and political economist Harold Innis (Mosco 1996). Innis's research about the development of the Canadian Pacific Railway, and later the role of commodities in establishing settlements in the Canadian hinterland, would form the basis for his staples theory, sometimes known as "dirt research." Staples, such as fur, fish, or lumber, dictated not only economic realities for regions and communities; they underpinned political and cultural life. Furthermore, each staple or resource contained a set of space-time biases, described by Barnes (2005) as "spatial and temporal imperatives that are then manifest once staples extraction and trade begins." Applying the framework to ancient civilizations, Innis argued that space-time imbalances could precipitate conflict and inequality (Innis 1948).

Innis's original interest in resource commodities such as pulp and paper eventually led him to analyse both newsprint and the newspaper industry, in turn moving his interest towards the study of media and technology. In particular, he was interested in the impact of such technologies on institutions and social structures. His focus remained on the form and structure of media and technologies rather than content (Patterson 1990)—including their spatial or time-based attributes. Technological media—ranging from papyrus scrolls to printed books, and telephony to radio—have tended to historically favour one dimension over the other (Barnes 2005). Radio, for example, with its ability to transmit a uniform signal across a vast geography, features a spatial bias.

Spatialization, then, allows media enterprises to adjoin previously separated audiences and geographies. The parallels between global media and transportation systems are noteworthy. Both transportation and communication minimize the time it takes "to move goods, people and messages over space, thereby diminishing the significance of spatial distance as a constraint on the expansion of capital markets" (Mosco 1996, p. 173). The role of spatial media in propping up institutions, including governments and corporations, was pervasive in Innis's work. At the same time, Innis recognized the disruptive qualities of emergent, competing media technologies, and their ability to challenge existing hierarchies or systems with counter-discourses.

Transformations in the global economy and within modern capitalism, and the changes in conceptions of space and time that have accompanied them (Harvey 1990), have ensured the durability of Innis's original thesis. Because drones are designed to overcome altitudinal and horizontal distances, their spatial qualities become readily evidenced. Simultaneously, physical spaces can be experienced materially and symbolically. News reporting therefore emerges as a practice of representation, constructing the meaning of places, including countries, regions, or communities (Gasher 2017). Having emerged as a crossroads of globalization, commodity extraction, and environmental protection/degradation, Canada's West Coast presents an appropriate site for case study. Vancouver Island, as the largest island on North America's West Coast, has featured prominently in environmental contestations involving resources extraction, and in particular the logging of old-growth forests.

Case Study: Activism over Environmental Space

In what follows, an illustrative case of using a drone technology to create a mediated environmental activism is provided, drawing from the campaign by the Ancient Forest Alliance in British Columbia, Canada. The author retrieved drone-produced videos from the video sharing service of YouTube, specifically from the AFA's public channel. Public relations and advocacy texts were also gathered from the organization's website. Finally, media texts highlighting the campaign were retrieved from the LexisNexis database, using the search terms "Central Walbran Valley" and "drone."

Within this case, I firstly illustrate how drone technology can be deployed to engage supporters and online audiences by way of social media. Secondly, I show how such visual content is further analysed and shared through traditional media channels. Importantly, the aim here is not to present drone technology as a panacea for environmental activism mediation needs. However, environmental communicators considering new means to mediating policy differences with governments, companies, and other adversaries can learn a great deal from early adapters integrating spatial technologies into their strategic activism repertoire.

British Columbia's Walbran Valley Conflict

Canada's most westerly province of British Columbia is home to 25 million hectares of old-growth forests, 4.5 million of which are protected by government agencies. Located near Port Alberni, British Columbia, on Vancouver Island, the Central Walbran is considered one of Canada's most impressive old-growth rainforests. It is also one of the largest unprotected old-growth forests on southern Vancouver Island. About 5500 hectares of the Lower Walbran were included in the Carmanah-Walbran Provincial Park in 1994. A further 7500 hectares in the Central and Upper Walbran Valleys were left unprotected. The Central Walbran is a popular destination for recreation and outdoors enthusiasts, including hikers, campers, anglers, hunters, and mushroom pickers.

The area also takes on historical significance for regional environmentalists. The valley was considered "ground zero" for opposition to logging interests because of its bounty of western red cedar, Sitka spruce, and hemlock trees—and early Walbran protests helped to build a broader momentum for the much larger Clayoquot Sound protests of 1993, which occurred further north on Vancouver Island near Tofino (Ancient Forest Alliance 2015).

During September of 2015, a logging permit was issued by the British Columbia government to the Canadian forestry company Teal-Jones for the first of eight cutblocks in the forest (a cutblock is a parcel of forest slated for harvest). Due to the area's rugged and often inaccessible terrain, helicopter-based logging, also known as heli-logging, was slated for several cutblocks in the valley, including the first one approved by the British Columbia Forest Service. Ken Wu, the executive director of the Ancient Forest Alliance, connected the logging projects to broader ecological concerns on Vancouver Island:

> The one thing the BC government must not do is to reward unsustainable practices with more unsustainable practices—but that's just what they've done by granting the first cutting permit to Teal-Jones in the Central Walbran Valley. It's a myopic government facilitating the demise of an ecosystem for a company intent to go just about to the very end. Instead they need a quick transition or exit strategy to get completely out of our last

> ancient forests and into a sustainable, value-added, second-growth forest industry. (Ancient Forest Alliance 2015, para. 7)

In a bid to save the old-growth rainforest from destruction at the hands of logging, West Coast conservationists who are part of the Ancient Forest Alliance (AFA) made the decision to employ remotely piloted drones. TJ Watt, an activist and photographer with the AFA, purchased a remote-controlled drone equipped with a GoPro camera—a DJI Phantom 2 w/ H3-3D gimbal and GoPro Hero 4 Black. His camera investment of just over $1000 allowed him to shoot images of massive trees in previously inaccessible sections of the forest due to steep and challenging terrain. The ability to transmit images of such wilderness had a twofold effect, according to Watt. Firstly, it allowed the organization to "get some spectacular footage of magnificent but endangered old-growth forests from vantage points rarely seen." Secondly, it raised the public's awareness of remote endangered areas that are normally out of public view, and in turn where companies might believe they can conduct operations with little scrutiny.

"Drones are a new tool in our tool box because for many people these trees might as well be on the moon," Watt said in an interview with *The Province*, a daily newspaper in Vancouver, BC. "They were out of sight and mind for most. But the drones let us raise environmental awareness about these remote endangered areas where companies believe they can log with little scrutiny. If we can't bring B.C.'s four million people to the forests, we're going to bring the forests to the people."

The organization's first video, released in 2015, was entitled "Save the Central Walbran Valley—Canada's Grandest Ancient Forest at Risk." It featured both aerial and ground-level footage of the 500 hectare Central Walbran Valley. This includes striking imagery of an area called Castle Grove, a clustering of ancient western redcedar trees. Other footage highlighted rivers, ponds, as well as aerial vistas of the valley. The video also included people interfacing with the forest and the trees—placing humankind within nature but also demonstrating the sheer scale and size of the trees. It denoted the recreational opportunities in the area, including hiking, fishing, camping, and mushroom picking. The three-minute video, featuring a soundtrack of ambient music, has garnered over 15,000 views since its publication.

A second video, released in the summer of 2016, highlighted Canada's second-largest Douglas-fir tree, dubbed "Big Lonely Doug." The 217-foot-high tree is located near the Central Walbran Valley and the

Vancouver Island coastal community of Port Renfrew. It sits in the middle of a large tract of forest that had been felled by industrial clearcutting of old-growth forest. The image of Big Lonely Doug standing by itself is juxtaposed by the stark landscape surrounding it. The video closes with a message that old-growth forests are key to the protection of endangered wildlife, water, and climate stability, but also First Nations' culture and local tourism. It concludes with a call to action—to sign the Ancient Forest Alliance's petition calling for a halt to the logging.

Analysis

Like their counterparts in the field of journalism, environmental activists use drones for their immediate applications such as reporting, monitoring, and image capture in remote or distant areas. These variables account for at least part of the drone's rise as a tool for environmentalists, engineers, and other occupations situated in remote or difficult geographies. Yet environmentalists also concede that for their audiences, nature can be an abstraction, one that may shield publics from the material reality of detrimental practices such as clearcut logging. The AFA's spokesperson argued that the contested old-growth trees "might as well be on the moon," in turn giving companies a free pass to log at will. Drones, then, provide conservationists with an important new tool in the battle to protect endangered wilderness areas. They illuminate the material conditions within otherwise abstract environmental debates, thus rendering their persuasion with a heightened sensory experience and additional layers of geographic detail. They help environmental communicators construct meaning within environmental debates, and they counter prevailing discourses produced by government or corporate bodies.

The Ancient Forest Alliance's decision to employ remotely piloted drones was but one tool within their arsenal of strategic communication tactics, complementing existing press releases, photography, and social media messages. Yet the material richness of this video, providing relativity and contextual detail to the mediation of contested wilderness areas, served to both deepen the environmental message and expand its footprint with a larger audience. Drone videography is capable of providing unparalleled visual detail and perspective of natural spaces. At the same time, the advent of social networking and video streaming venues such as Facebook, YouTube, and Vimeo allowed for efficient and interactive mediation of this emerging genre of activism. The activation of media relations techniques such as press releases further enabled the compelling aerial footage of the Central Walbran Valley to be featured in metropolitan and regional media—thus furthering the disruptive, challenging nature of this counter-discourse. Within this strategic communication, drones did not just help to produce a story; concurrent with Howley's (2018) assertion, they became the subject of the narrative as well.

Discussion

This chapter set out to explore the emerging interplay between technology, media, and ecology that the rise of drone media activism in the public sphere represents. AFA's drone-captured video of pristine rainforest in the remote Walbran Valley of Vancouver Island represents a new approach to monitoring, documenting, and publicizing both inaccessible geographies and the ecological battles they are immersed in. Drawing from the conceptualization of spatial communication developed by political economist and historian Harold Innis, the drone—when equipped with videography technology—emerges as a space-biased communication medium. While ownership of this technology was previously relegated to well-resourced institutions, which most notably deployed the technology for the purposes of warfare, surveillance, and monitoring, the increasing affordability of this technology has altered discourses and understandings of UAVs within environmentalism and media production. Activists are now in a position to monitor aerial extraction practices—including heli-logging—through aerial media capture devices of their own. Furthermore, captured video can be repackaged as public interest communication, bolstering the general public's understanding of regional, national, and international ecosystems.

Governmental and corporate institutions have utilized (and continue to use) drone technologies in cases where unmanned aircraft systems are deemed necessary for military, governance, or economic objectives. However, the more recent uptake of drone technology by environmental activists reveals a shift in how such a technology can be leveraged in the future—as a conduit for ecological information and meaning-making in the public sphere. It also underscores a tenet of Innis's theory of time- and space-media—that access to media technologies presents a means to propping up power structures or, conversely, disrupting them.

There is a potential for growing conflicts over the rights to use airspace and air transport technology as a vehicle for visual media capture, especially when high-profile companies and projects are involved. A case in point comes from the April 2016 leakage of the Keystone pipeline in South Dakota, when the anti-pipeline environmental group Bold Nebraska hired a pilot to surveil the scene. TransCanada, the owner of the pipeline, requested a no-fly order over the pipeline spill site (Hult 2016). The company argued that its request was made to secure the airspace for surveillance by clean-up crews, though this view was not shared by Bold Nebraska.

This example underscores one of the most significant challenges facing drone technology today: its operation in regulated, non-segregated airspace (Pereira et al. 2009). In the United States, legal obstacles will continue to be an overriding concern for drone operators. This is because commercial users must be licensed by the Federal Aviation Authority. More importantly, privacy laws preclude activists from flying over private property to acquire footage, which is considered "extraordinary means." Activists may work around this rule by shielding their efforts as journalism in the public interest, but this would be decided by the courts.

The legal questions that have been raised by UAVs in the realm of environmental advocacy raise new questions about the journalistic implications of mediated activism such as drone videography. Activists may argue that their communication, while in the public interest, is not afforded the same protections as what typically qualifies as journalism. This is likely to create a motivation for some environmental organizations to strategically position their public interest communication as journalism, as opposed to organizational advocacy or public relations. Large environmental organizations such as Greenpeace and the Sierra Club are already moving in this direction, publishing their stories and blog posts on in-house-produced news websites.

Beyond legal implications, the aerial nature of drone videography ensures that it will be well represented within mass media, including news but also popular culture. The usage of drones by environmentalists aligns with a broader interest in aviation-based popular communication. For example, the television programme *Aerial America* has emerged as a franchise programme for the Smithsonian Channel cable network, providing sweeping geographic overviews of the topographies and landscapes of each US state. While the programme offers a superficial but visually impressive tour of major landmarks—geological, historic, urban, cultural—it has also provided new perspectives on environmental debates and disasters, such as mountaintop removal coal mining in West Virginia's Appalachian Mountains. Like-minded programmes based on aerial video footage, ostensibly serving armchair tourists, have weaved key ecological themes into their programme scripts, simply by showing topographies and landscapes otherwise left unseen. The phenomenon of hidden ecology being mediated through the mass media thanks to technological advances conjures up Mann and Twiss's (1910) eco-philosophical adage. When environmental events are made visible to the public, their legacy is secured. The drone's ability to conquer physical space and document the previously unseen—or the otherwise unknown—not only reshapes contemporary debates about the natural world. It recalibrates society's relationship with the planet—spatially articulating the global ecological crisis simultaneously through material and symbolic lenses.

Acknowledgements The author thanks Dan Morrison from the University of Oregon's School of Journalism and Communication for his insights into this topic.

References

Ancient Forest Alliance. (2015, September 22). Drones used in BC's Old-Growth Forest Campaigns – Walbran Valley conflict escalates. Retrieved from http://www.ancientforestalliance.org/news-item.php?ID=922

Barnes, T. (2005). Borderline communities-Canadian single industry towns, staples, and Harold Innis. In H. van Houtum (Ed.), *B/ordering space* (pp. 109–122). Routledge: Abingdon.

Blair, J. A. (2012). The rhetoric of visual arguments. In *Groundwork in the theory of argumentation* (pp. 261–279). Dordrecht: Springer.

Carpentier, N. (2017). *The discursive-material knot. Cyprus in conflict and community media participation*. New York: Peter Lang.

Charles, A. (2016). Dogs. In *Political animals: News of the natural world* (pp. 155–176). Oxford: Peter Lang.

Choi-Fitzpatrick, A. (2014). Drones for good: Technological innovations, social movements, and the state. *Journal of International Affairs, 68*(1), 19.

Conniff, R. (2017). Eyes on nature: How satellite imagery is transforming conservation science. Yale Environment 360. Retrieved from https://e360.yale.edu/features/eyes-on-nature-how-satellite-imagery-is-transforming-conservation-science

Corcoran, M. (2014). *Drone journalism: Newsgathering applications of Unmanned Aerial Vehicles (UAVs) in covering conflict, civil unrest and disaster*. Adelaide: Flinders University.

Culver, K. B. (2014). From battlefield to newsroom: Ethical implications of drone technology in journalism. *Journal of Mass Media Ethics, 29*(1), 52–64.

DeLuca, K. M. (2005). *Image politics: The new rhetoric of environmental activism*. Abingdon: Psychology Press.

Franke, U. E. (2015a). Drone wars: Transforming conflict, law, and policy. *Parameters, 45*(2), 128–131.

Franke, U. E. (2015b, January 26). Civilian drones: Fixing an image problem?. *ISN Blog. International Relations and Security Network*. Retrieved March 5, 2015. https://isnblog.ethz.ch/security/civilian-drones-fixing-an-image-problem

Gasher, M. (2017). Sustaining spaces, patrolling borders: News coverage of migrating peoples. In *What is sustainable journalism?* (pp. 181–198). Oxford: Peter Lang.

Gilliam, F. D., Jr., & Bales, S. N. (2002). Strategic frame analysis and youth development: How communications research engages the public. In *Handbook of applied developmental science: Applying developmental science for youth and families: Historical and theoretical foundations*. Thousand Oaks: Sage.

Gynnild, A. (2014). Journalism innovation leads to innovation journalism: The impact of computational exploration on changing mindsets. *Journalism, 15*(6), 713–730.

Harvey, D. (1990). Between space and time: reflections on the geographical imagination. *Annals of the Association of American Geographers, 80*(3), 418–434.

Heimbuch, J. (2011). Activists launch drones to battle Japanese whalers. *Treehugger*. Retrieved from https://www.treehugger.com/gadgets/activists-launch-drones-to-battle-japanese-whalers.html

Hill, C. A. (2012). The psychology of rhetorical images. In *Defining visual rhetorics* (pp. 37–52). Routledge: Mahwah, New Jersey.

Holton, A. E., Lawson, S., & Love, C. (2015). Unmanned aerial vehicles: Opportunities, barriers, and the future of "drone journalism". *Journalism Practice, 9*(5), 634–650.

Howley, K. (2018). *Drones: Media discourse and the public imagination*. Oxford: Peter Lang.

Hult, J. (2016). FAA: TransCanada requested no-fly zone. Retrieved April 11, 2016, from http://www.argusleader.com/story/news/2016/04/06/faa-transcanadarequested-no-fly-zone/82714246/

Kitchin, R., Lauriault, T. P., & Wilson, M. W. (Eds.). (2017). *Understanding spatial media*. Sage: London.

Koh, L. P., & Wich, S. A. (2012). Dawn of drone ecology: low-cost autonomous aerial vehicles for conservation. *Tropical Conservation Science, 5*(2), 121–132.

Lefebvre, H., & Enders, M. J. (1976). Reflections on the politics of space. *Antipode, 8*(2), 30–37.

Lester, L., & Hutchins, B. (2012). The power of the unseen: Environmental conflict, the media and invisibility. *Media, Culture & Society, 34*(7), 847–863.

Love, C. D., Lawson, S. T., & Holton, A. E. (2015). News from above: First amendment implications of the Federal Aviation Administration ban on commercial drones. *Boston University Journal of Science & Technology Law, 21*, 22.

Mann, C. R., & Twiss, G. R. (1910). *Physics*. Chicago: Scott, Foresman, & Company.

Moore, A. (2015, August 12). Precious turtle eggs stolen from Mexican beach. *Reuters*. Retrieved from https://www.reuters.com/video/2015/08/12/precious-turtle-eggs-stolen-from-mexican?videoId=365269053&videoChannel=4

Mosco, V. (1996). *The political economy of communication: Rethinking and renewal* (Vol. 13). London: Sage.

Patterson, G. (1990). *History and communications: Harold Innis, Marshall McLuhan: The interpretation of history*. Toronto: University of Toronto Press.

Pereira, E., Bencatel, R., Correia, J., Félix, L., Gonçalves, G., Morgado, J., & Sousa, J. (2009). Unmanned air vehicles for coastal and environmental research. *Journal of Coastal Research, 2*, 1557–1561.

Phadke, R. (2010). Defending place in the google earth age. *Ethics Place and Environment (Ethics, Place & Environment (Merged with Philosophy and Geography)), 13*(3), 267–281.

Radjawali, I., & Pye, O. (2017). Drones for justice: Inclusive technology and river-related action research along the Kapuas. *Geographica Helvetica, 72*(1), 17.

Sandbrook, C. (2015). The social implications of using drones for biodiversity conservation. *Ambio, 44*(4), 636–647.

Sidik, A. (2012). Meat packing plant under investigation for dumping pig blood into nearby creek. *GreenAnswers.com*. http://greenanswers.com/news/274701/meat-packing-plant-under-investigation-dumping-pig-blood-nearby-creek

Simpson, E. (2015). Nothing monotonous about drones now. *Political Science Publications*. Retrieved from https://ir.lib.uwo.ca/cgi/viewcontent.cgi?article=1096&context=politicalsciencepub

Tremayne, M., & Clark, A. (2014). New perspectives from the sky: Unmanned aerial vehicles and journalism. *Digital Journalism, 2*(2), 232–246.

Zenko, M. (2012). 10 things you didn't know about drones. *Foreign Policy*. Retrieved from https://foreignpolicy.com/2012/02/27/10-things-you-didnt-know-about-drones/

Zook, M. A., & Graham, M. (2007). Mapping DigiPlace: Geocoded Internet data and the representation of place. *Environment and Planning B: Planning and Design, 34*(3), 466–482.

Employing Digital Technologies in Environmental Communication Campaigns in China: A Capacity-Building Perspective

Kenneth C. C. Yang and Yowei Kang

Introduction

China's rapid ascent to a global economic superpower has mainly depended on its cost-competitive manufacturing capabilities (Kamrany and Jiang 2015; The Economist 2015). Ironically, China's role as "the world factory" has impacted her already fragile environment (Liu 2014; Sangay 2017; The Economist 2013) after lowering her own environmental criteria to attract foreign business ventures (Albert and Xu 2016). In November

K. C. C. Yang (✉)
Department of Communication, The University of Texas at El Paso,
El Paso, TX, USA
e-mail: cyang@utep.edu

Y. Kang
National Taiwan Ocean University, Keelung City, Taiwan
e-mail: yoweikang@mail.ntou.edu.tw

J. Díaz-Pont et al. (eds.), *The Local and the Digital in Environmental Communication*, Global Transformations in Media and Communication Research - A Palgrave and IAMCR Series,
https://doi.org/10.1007/978-3-030-37330-6_3

2010, China reluctantly admitted that it was the world's largest emitter of greenhouse gases (Institute of Energy Research 2015). *The Economist* (2013) similarly reports that China is the world's largest polluter and $275 billion has been allocated to address the air pollution problem rampant in China's major cities.

Both local and foreign-affiliated environmental non-government organizations (henceforth, ENGOs) in China have played significant roles in protecting her fragile ecosystem. However, many Chinese ENGOs are operated by private citizens of her nascent civil society and are often constrained by their organizational capacity (Tang and Zhan 2011). China's ENGOs are found to "still struggle for support" from many stakeholders at both local and national levels (Sha 2012, n.p.). Chinese ENGOs have a long tradition of surviving in a difficult political context through the creation of "innovative organization and practices" (Hsu et al. 2016). Therefore, we posit that digital technologies function as an innovative practice to cultivate ENGOs' organizational capacity effectively through the mobilization of local resources in shaping government policies.

In recent years, under President Xi's authoritarian helm, the Chinese Communist Party (henceforth, CCP) government resumes its full control over whatever resources might exist in society (Wong 2016), further worsening local ENGOs' capabilities to create environmental communication campaigns to shape local stakeholders' attitude and behaviour changes. In particular, the passage of the *2016 PRC Law on the Management of Foreign NGOs* has affected many foreign-affiliated ENGOs in their capacity-building activities, ranging from fundraising, volunteer recruitment, and other operations, for example, partnerships with other local NGOs in this country (Wong 2016). As a result, whether or not ENGOs are able to integrate digital technologies into their capacity building affects not only their growth and survival (Gao 2013), but also the effectiveness of their environmental communication campaigns.

China has presented an interesting case to demonstrate the constant contestations and dynamic interactions among government policies, ENGOs' conservation objectives, capabilities of digital technologies, and her political setting to influence environmental communication at the local level. A comprehensive understanding of China's political context is crucial for the successful implementation of these digital technologies to enable ENGOs to think globally to accomplish the common good for human beings, but act/adapt locally to the social and political realities in China. In other words, despite China having risen as an indispensable manufacturer in the global economy, her own highly localized environmental policies have failed to meet international standards. Consequently, air pollution is estimated to reduce the life expectancy of Chinese living in northern China by about 5½ years, according to a study by America's National Academy of Science (The Economist 2013). As a country that claims to own the largest network of high-speed bullet trains (Jacobs 2019) and sophisticated surveillance cameras with artificial intelligence capabilities (Chan 2018), China bans *Facebook*, *YouTube*, *Instagram*, and many major Western social media, yet chooses to develop her localized versions of these technologies.

These paradoxes offer environmental communication scholars an excellent occasion to observe how China's locality characteristics could affect the opportunities brought up by the emerging digital technologies. This chapter investigates the local impacts on the technology-enabled capacity-building practices of several Chinese ENGOs in order to enhance environmental activism, inter-ENGO collaboration, and local participation. This chapter is written to address two organizing themes in this edited volume: (1) to examine the roles of digital technologies in ENGOs' capacity building by mobilizing grassroots activism, networking, and participation at the local level and (2) to demonstrate the approaches Chinese ENGOs have taken to localize technology applications (such as the inclusion of QR Code) in their environmental communication campaigns.

Literature Review and Theoretical Framework

This chapter aims to examine the roles of globally diffused mobile social media in relation to ENGOs' capacity building in China. Past literature on non-profit organizations and their use of digital technologies provides rich description, but often ignores in what ways these technologies may help non-profits with a variety of their organizational functions (see Guo and Saxton 2014a for review). The following literature review is divided into two sections. First, we provide an overview of digital technology developments in China. Second, we discuss how these digital technologies are instrumental in building ENGOs' organizational capabilities in China to develop more effective environmental communication campaigns.

The Development of Digital Technologies in China

Digital technologies broadly refer to computer, the Internet, mobile technologies, and other converging technologies including the above (Temple University 2017). China's mobile-cellular penetration is very high (93.16% per 100 inhabitants; International Telecommunication Union 2016). There are approximately 0.64 billion smartphone users and the number is expected to grow to 0.69 billion in China (eMarketer 2015). According to the *Statistical Report on Internet Development in China*, 95.1% of Internet users also use a mobile device to access the Internet in 2016, rising from 24.0% in 2007 (eMarketer 2017). Mobile Internet users have grown from 87.5 million subscribers in 2014 to 109.4 million users in 2016 (National Bureau of Statistics of China 2018). Furthermore, social media penetration in China is 64%, comparable to other countries in this region: South Korea (84%), Taiwan (80%), Singapore (79%), Hong Kong (76%), Macau (66%), and Japan (55%) (Kepios 2017).

Digital technologies are often speculated to influence how members of a public perceive and engage with environmental communication problems (Arts et al. 2015). Exponential penetration statistics above suggest that digital technologies, like mass media, can offer local ENGOs opportunities to increase people's awareness, knowledge, and behaviours (Saleh and Saifudin 2017). For the organizational capacity-building perspective, these technologies are known for their social mobilization capacities (Howard and Hussain 2011) to help resource-deficient ENGOs in China. Among many technological innovations, the emergence of social media has apparently transformed China politically, culturally, and socially (Hong 2017; Shao and Wang 2017) because of the less stringent control than state-owned traditional media. Despite the lack of *Facebook*, *YouTube*, *Foursquare*, and *Twitter*, China's social media landscape, such as privately owned *Baidu*, *Tencent*, *Alibaba*, *Weibo*, *Douban*, *RenRen*, *Baidu TieBa*, and so on, is still burgeoning in parallel with its Western counterparts (Chiu et al. 2012). The development of social media in China demonstrates their highly localized characteristics, due to the restrictions by the Chinese government that ban foreign social media from operating in China (Price and Reuters 2018). Therefore, we bring up our first research question: What approaches have Chinese ENGOs taken in terms of using digital technologies to enhance environmental activism, stakeholder engagement, inter-ENGO collaboration, and local stakeholder participation in their environmental communication campaigns?

The Roles of Digital Technologies and ENGOs' Capacity Building

Extant research on the impacts of digital technologies on environmental communication has examined the functions, behaviours, and strategies of ENGOs, but has not associated digital technologies with organizational capacity building. For example, Arts et al. (2015) identify five dimensions of digital technology impacts on nature conservation (i.e., communication and experience, data on both people and nature, data integration and analysis, and participatory governance). Although these dimensions are related to ENGOs' capabilities, they have used the term "capacity building" in their analysis. Similarly, Guo and Saxton (2014a, b) examine the role of Twitter in policy advocacy, but technology-enabled capacity building is not the main concern of their studies. To fill the gap in the current literature, this chapter is based on an organizational capacity-building perspective to examine the roles of digital technologies in local environmental communication campaigns.

Capacity building refers to "the ability of nonprofit organizations to fulfill their missions in an effective manner" (McPhee and Bare 2001, p. 6). The term often refers to a long-term process that helps facilitate "knowledge acquisition to individuals about a particular sector, so that they can get empowered and acquire appropriate skills, to successfully perform their required functions, towards improving the quality of their life" (Zeitoun 2003, n.p.). The study of the organizational capacity-building process often involves the examination of how ENGOs develop and mobilize resources, foster specialized skills, establish management systems, set up learning networks, and encourage the fulfilment of organizations' missions and viable development (Büscher 2014; Li and Guo 2015; NPO Information Consultation Center 2003).

Recent advances in digital technologies have generated interest among environmental communication researchers and practitioners to examine their impacts on local campaigns. For example, researchers have acknowledged and examined the roles of digital technologies in NGOs' capacity building (Büscher 2014; De Vita et al. 2001; Gevaert 2012; UNESCO n.d.). Digital technologies have also helped *Friends of Nature* in China to develop campaigns to address local soil pollution problems. *Friends of Nature* engages local stakeholders through mobile social media to take part in its public interest environmental lawsuit against a major mining operator and soil polluter in Nanjing (Chun 2015). *Friends of Nature* uses these technologies to recruit class action plaintiffs, donors, volunteers, local pollution monitors from the polluted area to help enact government's environmental policies and law enforcement (Wilson 2015). While the mobilization of public participation through digital technologies is a success to create public awareness of this case, the lawsuit against the illegal quarry operator turns out to have a mixed result (Wilson 2015). While the court has imposed heavy fines on the mining company, the CCP government is wary about any environmental activism in China because, once a large number of plaintiffs are mobilized, the local government will be held liable for failing to enforce environmental regulations (Wilson 2015).

The above example has shown that the potential of technologies to encourage participation at the local level can be constrained by non-technological factors (such as the CCP government's concern over mass protest and mobilization in an authoritarian regime). This example prompts environmental scholars to rethink their technology determinism perspective that has only focused on the relationships between characteristics of digital technologies and ENGOs' activities. Extant research is inherently prejudiced because its deterministic perspective assumes favourable technology characteristics will help ENGOs, without considering external environmental factors. One of the optimistic claims is Kaplan's (2012) that identifies four I's of mobile social media through which organizations will be able to individualize (I) and integrate (I) their activities to create relevance in participants' daily life. Scholars who follow Kaplan's (2012) line of reasoning often argue that these highly engaging ENGO activities through mobile social media will result in the involvement (I) of users to initiate (I) the content-creation process to assist ENGOs' capacity building. Other examples include Büscher (2014) that observes that Web 2.0 and social media applications are crucial for ENGOs to reach their target audiences through the sharing, co-creation, and assessment of website contents. Similarly, Good (2006) reviews and summarizes the role of the Internet in resource management and environment activism for ENGOs. Good (2006) concludes that the "Internet facilitated speed, cost-effectiveness, interactivity, freedom of expression, and global reach for environmental organizations" (p. 10).

Despite the enthusiastic descriptions of digital technologies to transform environmental communication, existing literature has yet to explore whether and how locality factors might limit whether these potentials can be fully realized in China. Actual applications of digital technologies often result in mixed outcomes in generating intended outcomes at the local level. For example, *JGI China*, a local branch of Jane Goodall Institute, has used *Taobao's Online Charity Store* to raise funds; however, this application is still plagued by the lack of trust and credibility on NGOs, the absence of customer service personnel, and the lukewarm responses from Chinese online shoppers (China Development Brief 2013). The above examples demonstrate whether digital technologies that will help Chinese ENGOs to build their organizational capacities are influenced by political-social-cultural factors in China.

Our analytical framework is based on De Vita et al.'s (2001), which touches upon the interactions among various organizational characteristics in the capacity-building processes (De Vita et al. 2001). To develop a technology-centric model, we also integrate Kaplan's (2012) four I's because of their close relationships with digital technologies discussed in this chapter. To address the gap in the existing capacity-building literature for ENGOs, we further include external environmental factors to account for the impacts of locality determinants on digital media applications. We reason that this framework will take into consideration the locality factors when explaining how digital technologies influence ENGOs' capacity building. In this model, the core elements of ENGOs' capacity building are composed of their vision, mission, and leadership that interact with organizational resource management, outreach activities, and conservation products/services. Mobile social media and their technological characteristics (i.e., Kaplan's four I's) are posited to conceptualize their influence on ENGOs' capacity-building functions. Cultural, political, social, and regulatory factors as external environmental determinants are posited to influence how and to what extent ENGOs employ digital technologies to build their capacities (Fig. 1).

Based on the theoretical model, our study aims to answer the following research questions:

RQ1: What approaches have Chinese ENGOs taken in terms of using technologies to enhance environmental activism, stakeholder engagement, inter-ENGO collaboration, and local stakeholder participation in their environmental communication campaigns?

RQ2: Will China's political-social-cultural factors explain how Chinese ENGOs use technologies to build their outreach and resource mobilization capabilities to develop more effective local environmental communication campaigns?

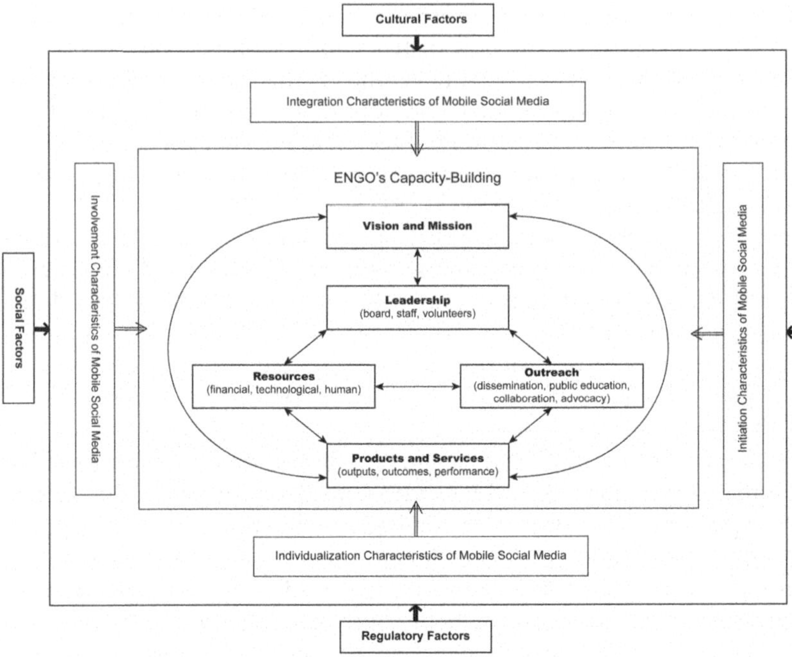

Fig. 1 Our theoretical model

Research Method

The Selection of Case Study Method

Using a case study approach (Francesch-Huidobro and Mai 2012; Li et al. 2017), we chose several Chinese ENGOs (e.g., *Friends of Nature*, *WildAid China*, and *Conservation International China Program*) that demonstrate the best practices in using digital technologies to build their organizational capacity through effective local environmental communication campaigns. *Conservation International China Program (CI)* operates as a local branch of *Conservation International*, a global conservation organization with over 200 partners in 30 countries. On the other hand, *WildAid China* is also a local branch office of *WildAid* that operates in Ecuador, India, Indonesia, the UK, and the US.

The selection of these two ENGOs in this study is based on the list compiled by United States Environmental Directories, Inc. (http://www.earthdirectory.net/china#china), which includes over 80 ENGOs currently operating in China. We visited their websites to identify and select the ENGOs with best digital technology applications relevant to the themes of this edited book to examine the influence of digital technologies on local environmental communication campaigns.

Discussion

The following discussions are divided into two sections to answer our research questions. First, we discuss the approaches that two local ENGOs have taken to employ mobile social technologies in their environmental communication campaigns. We observe that these two ENGOs have adapted their environmental communication campaign to select the most popular technological applications in China. For example, the *WildAid China* website has included QR Code to provide access to its localized technologies such as *Weibo* (https://wildaidchina.org/), while its global website outside China (https://wildaid.org/about/stay-in-touch/) lists only Western social media technologies (such as *Google+*, *Instagram*, *Twitter*, *Pinterest*, *YouTube*, and *LinkedIn*). This has supported our previous argument that the implementation of digital technologies depends on a country's external environment where China has completely banned foreign social media. These Chinese ENGOs have adopted different mobile social media to develop their capacity as demonstrated in these campaigns. In the following subsections, we also examine how mobile social technologies are adopted by local Chinese ENGOs to build their organizational capabilities to develop more effective local environmental communication campaigns.

Employing Mobile Social Media for ENGOs' Capacity-Building Activities

Among many capacity-building activities, we observe that Chinese ENGOs use digital technologies to share information among and with their stakeholders (Yang 2005). Mobile social media allow Chinese ENGOs to spread news about environmental problems to local residents and other ENGOs cost-effectively. For example, the establishment of a virtual collaborative network with *WildAid*, *Friends of Nature*, and other ENGOs in

China allows traditional media to become fully aware of their anti-pollution campaign (Jiang 2017). The mobile social media site (with several QR Codes embedded) has successfully influenced the government's policies to postpone the construction of a large hydro-plant project that several ENGOs claim will endanger the habitat of rare peacocks (http://www.fon.org.cn/index.php?option=com_k2&view=item&id=12833&Itemid=121).

Information-sharing is an important part of ENGOs' outreach activities to build their capacity. *WildAid* has used mobile social media to create public awareness, engage citizens, and encourage civil society to address conservation problems in China. As shown above, several Chinese ENGOs use mobile social media to share information about pending constructions that may have irrevocable impacts on the habitats of rare peacocks. The information-sharing capabilities are important for Chinese ENGOs where the state-owned media often censor environmental news and prevent local ENGOs from using traditional media to inform and mobilize local stakeholders to take part in petitioning and mass protesting (Forbes 2015). Mobile social media has also become a viable option to bypass government censorship and mobilize 1000 people to join the mass protest against the construction of a chemical plant in Shanghai (Forbes 2015). The large number of local participants offers ENGOs a bargaining chip with the government that often succumbs to the public pressure to propose a short-term middle-ground solution to appease to the situation (Forbes 2015). As shown in the above case, mobile social media are capable of mobilizing local stakeholders to draw attention from both the public and the government officials. This is similar to the concept of "attention philanthropy" that stresses the importance of obtaining a larger audience base through these outreach activities to communicate and engage with local stakeholders to receive attention (Guo and Saxton 2014b).

Localized technologies in China play a similar role as Western mobile social media in a variety of capacity-building functions. *WildAid* has chosen *Weibo* (similar to *Twitter* outside China) to reach out to its local stakeholders in China to build its capacity. Its main *Weibo* site attempts to warn Chinese consumers and tourists of the negative consequences of purchasing souvenirs made of ivory and tortoise shell. *WildAid* employs the social media platforms most popular among Chinese users to deliver messages that discourage the consumption of armadillo and shark fins. Its social media campaigns include celebrity endorsements (such as Jackie Chen, Jay Chou, and Yao Ming) to save endangered wildlife by never purchasing any sea turtle or pangolin products. Localized mobile social media (such as *Weibo* or *Weixin*) also enable ENGOs to reach out and communicate with their followers regularly to generate the public attention ENGOs need (Guo and Saxton 2014a). In addition, these emerging platforms also help the *WildAid* network and other ENGOs to successfully launch several lawsuits against major pollutants in China. Furthermore, the formation of a social media advocacy network has become an important part of capacity-building activities to reach out to other relevant stakeholders through information dissemination, inter-ENGO collaboration, and public education of the Chinese public (De Vita et al. 2001).

Another advantage of mobile social media technologies is that they can also easily be integrated with other platforms to influence attitude and behaviour changes among local stakeholders. For example, *WildAid*'s spin-off project, the *GoBlue* air pollution campaign (https://www.wildaidgoblue.com/), promotes a low-carbon and sustainable lifestyle by encouraging people to use public transportation or walk to work. The campaign website includes links to access other social media and mobile platforms: *Youku* (similar to *YOUTUBE*) for delivering several promotional public service announcement (PSA) videos, *Weibo* for combining consumer-generated contents with sweepstakes (https://m.weibo.cn/status/4284508246817488?), and QR Code for smooth access to campaign information. The campaign has accumulated over 250 million impressions with its PSA videos and 115 million impressions with the billboard platform across China (WildAid 2018). The campaign has successfully generated awareness among Chinese consumers to reduce carbon emission to alleviate climate change effects (WildAid 2018). Launched in 2016, *WildAid's Go Blue* campaign has reported a successful outcome; it has reported that 73% of car users have reduced car usage (WildAid China 2016).

Chinese ENGOs often employ digital technologies available only to Chinese ENGOs. However, both CI and *WildAid* have adopted strategies similar to those outside China; that is, mobile social media are used to involve (I) interested participants by initiating (I) the creation of their own contents to describe how these ENGOs and their environmental communication campaigns are related to stakeholders' daily lives. Two of the four I's (i.e., Involvement and Initiation) (Kaplan 2012) are employed by both CI and *WildAid*. User-generated contents are more effective in engaging consumers (Kaplan 2012) and are more likely to help ENGOs recruit a larger volunteer base and contribute to the building of human resources as part of their capacity-building objectives.

We observe that QR Code emerges as one of the most popular mobile platforms among Chinese ENGOs. It is a mobile marketing tool that can be easily embedded in existing advertising messages (Ryu 2013). This mobile technology is irreplaceable in China because it has permeated people's lives, ranging from business transaction, government affairs, beggars requesting money, and brides receiving gift money (Wang 2017). From a capacity-building perspective, this mobile innovation helps ENGOs deliver their campaign messages in an effective manner in a country where smartphone penetration is expected to be over 70% in 2020 (eMarketer 2016). As a result, many Chinese ENGOs have adopted the QR Code technology to build their capacities to allow potential donors and volunteers to access their campaign site easily and smoothly.

QR Code also helps ENGOs' capacity building in terms of engaging with supporters and establishing a long-term relationship (Butler 2012) with the general public, building a larger base for support (De Vita et al. 2001) through volunteer recruitment and local community engagement. Furthermore, as an important tool in mobile marketing, QR Code is integrative (Kaplan 2012) in nature because it offers an easy link from mobile to ENGOs' Internet and social media platforms to increase involvement of the public. Some NGOs have used QR Code with their print ads to add access to audios, images, photos, videos, or other multimedia contents (Butler 2012). In terms of intra- and inter-organization communications, QR Code can also contribute to the sharing of ENGOs' vision and mission from the leadership to the entire staff within the organization and is instrumental in its organizational capacity building (Schneider 2014).

Both CI and *WildAid* have employed QR Code as an integrative and involving technology in their websites to inform local stakeholders, to engage and involve volunteers from local communities, or to encourage participation by watching videos on *Youku* (a Chinese version of *YouTube*). Widespread QR Code applications have demonstrated that the ascent of mobile social media in China prompts Chinese ENGOs to take a different path of technological applications from their parent organizations in the US in terms of recruiting volunteers, disseminating information, and building their networks among stakeholders. Although QR Code applications have the potential of fundraising to solicit donations to build financial capacity, as suggested by Western Non-Profit Organizations (NPO) gurus (Nonprofit Tech for Good 2011), Chinese ENGOs in this study do not include this function. The limited use of QR Code functionality is another example of how local regulatory and political factors may influence the applications of digital technologies to optimize their full potential in China.

According to China Development Brief (2013), only state-organized or "quasi-state run" charitable organizations can solicit online donations through a variety of platforms. Based on *The Law of the People's Republic of China on Public Welfare Donations*, only registered NGOs (under the categories of public welfare social organizations, non-profit public service units, and government departments) can solicit donations, offline or online, from the public (China Development Brief 2013). Registration with the Civil Affairs Department is usually time-consuming and requires a large start-up capital (China Development Brief 2013). The newly implemented *People's Republic of China Law on the Management of Foreign Non-Governmental Organizations' Activities Within Mainland China* (China Law Translate.com 2016) places similar emphasis on registration as its prerequisite to subsequent fundraising activities.

Conclusion

In conclusion, mobile social media technologies are instrumental in pro-conservation actions (Ballew et al. 2015). Given the increasingly unfriendly social and political milieu in China, the potential of mobile social media to generate local campaign impacts is likely to be limited. The objectives of this chapter aim to describe the approaches Chinese ENGOs have taken to integrate digital technologies into local environmental communication campaigns in China through fostering local environmental activism and participation and networking with local stakeholders and other ENGOs across China. For example, to complement its *GoBlue* campaign, *WildAid* has also launched its *Bike to Movies*, *Bike to Shop*, and *GoBlue EV* bus service events to increase the use of bikes and other low-carbon transportation in China's tier 1 cities (such as Beijing, Guangzhou, and Shanghai) (WildAid China 2016).

This chapter discusses innovative technological applications by several Chinese ENGOs that not only take advantage of characteristics of these mobile social media, but also develop effective local environmental communication campaigns after taking into consideration China's political-social-cultural factors. QR Code, as a capacity-building tool, is the most popular mobile technology among these ENGOs because of its already widespread adoption in China (Wang 2017). Although the extensive implementation of QR Code applications has been advocated in the literature from the West (Nonprofit Tech for Good 2011), its integration into several environmental communication campaigns by CI and *WildAid* is a result of China's technological development, as well as its unique political and social context. The so-called post-00s, -90s, and -80s generations are predicted to be the core users and consumers of many services and applications related to digital technologies (CIW Team 2018a, b). Post-90 mobile Internet users have reached 212 million as of February 2018 (CIW Team 2018a, b). The popular use of QR Code, in contrast to the complete absence of Western ENGOs, presents a strong case that locality determinants are likely to influence the success of a technology deployment among Chinese ENGOs.

The adoption of mobile social media among ENGOs aims to take advantage of their technological characteristics (De Vita et al. 2001; Guo and Saxton 2014a, b). For mobile social media, their capabilities to individualize (or personalize) messages, involve the public, initiate user content creation, and integrate multimedia and multiple platforms (Kaplan 2012) continue to be the major determinants impacting Chinese ENGOs' applications of these technologies. Nevertheless, using mobile social media to recruit volunteers or to mobilize local stakeholders for environmental causes is also affected by changing technological and political circumstances.

Limitations and Future Research Directions

This book chapter is constrained given the limited number of ENGOs that we have analysed. Furthermore, our analyses are based on their mobile social media applications to examine their roles in their environmental communication campaigns as well as organizational capacity building. Our approach is like Guo and Saxton's (2014a) that employs a website content analysis method to study how social media are used by advocacy non-profits in the US. This study, however, does not analyse message-level applications as Guo and Saxton (2014a) have done. Our findings should be interpreted with caution because we did not interview ENGOs' board members, staff, volunteers, and so on to solicit their viewpoints as to how mobile social media applications have helped their local environmental communication campaigns, quantitative outcome assessment, and other capacity-building efforts. Future research might shed more light on the decision-making process of digital technology adoption among ENGOs

(Nah and Saxton 2013) to examine what local impacts are intended with expensive investment in these technologies. Lastly, the rapid development of digital technologies also poses challenges to our study. For example, according to Nonprofit Tech for Good (2016), the rapid ascent of Internet of Things (IOT) has increasingly attracted many NGOs' attention to take advantage of its potential. Similar to the mobile social media discussed in this chapter, the success of IOT also depends on external environmental factors among individual countries where the invasion of privacy is acceptable. Concerns about rising government surveillance after the implementation of its social credit system may deter people from taking part in ENGOs' activities, deemed as "trouble-making."

References

Albert, E., & Xu, B. (2016, January 18). China's environmental crisis. *Council on Foreign Relations*. https://www.cfr.org/backgrounder/chinas-environmental-crisis. Accessed 30 July 2018.

Arts, K., van der Wal, R., & Adams, W. M. (2015). Digital technology and the conservation of nature. *Ambio, 44*(4), 661–673.

Ballew, M. T., Omoto, A. M., & Winter, P. L. (2015). Using Web 2.0 and social media technologies to foster pro-environmental action. *Sustainability, 7*, 10620–10648.

Büscher, B. (2014). Nature 2.0: Exploring and theorizing the links between new media and nature conservation. *New Media & Society, 18*, 726–743.

Butler, L. (2012, May 30). How charities can use QR codes. *The Guardian*. https://www.theguardian.com/voluntary-sector-network/2012/may/2030/using-qr-codes-charity-fundraising. Accessed 29 July 2018.

Chan, T. F. (2018, March 26). Parts of China are using facial recognition technology that can scan the country's entire population in one second. *Business Insider*. https://www.businessinsider.com/china-facial-recognition-technology-works-in-one-second-2018-2013/. Accessed 31 May 2019.

China Development Brief. (2013, October). Online fundraising strategies for grassroots NGOs. http://www.chinadevelopmentbrief.cn/articles/online-fundraising-strategies-for-grassroots-ngos/. Accessed 30 July 2018.

Chinalawtranslate.com. (2016, April 28). 2016 PRC law on the management of foreign non-governmental organizations' activities within Mainland China. https://www.chinalawtranslate.com/2016-foreign-ngo-law/?lang=en#_Toc449672032. Accessed 30 July 2018.

Chiu, C., Ip, C., & Silverman, A. (2012, April). Understanding social media in China. *McKinsey Quarterly*. https://www.mckinsey.com/business-functions/

marketing-and-sales/our-insights/understanding-social-media-in-china. Accessed 26 July 2018.

Chun, Z. (2015, November 14). NGOs win China's first public interest environmental lawsuit. *The Diplomat*. https://thediplomat.com/2015/2011/ngos-win-chinas-first-public-interest-environmental-lawsuit/. Accessed 1 June 2019.

CIW Team. (2018a, April 11). How does Tencent's QQ influence the young? *China Internet Watch*. https://www.chinainternetwatch.com/23668/qq-influence-social/. Accessed 26 July 2018.

CIW Team. (2018b, April 17). [REPORT] Post-90s mobile user study in lower-tier cities in China. *China Internet Watch*. https://www.chinainternetwatch.com/23723/post-23790s-lower-tier-cities-mobile-life/. Accessed 26 July 2018.

De Vita, C. J., Fleming, C., & Twombly, E. C. (2001). Building nonprofit capacity: A framework for addressing the problem. In C. J. De Vita & C. Fleming (Eds.), *Building capacity in nonprofit organizations* (pp. 5–30). Washington, DC: The Urban Institute.

eMarketer. (2015, September 16). Asia-Pacific boasts more than 1 billion smartphone users. https://www.emarketer.com/Article/Asia-Pacific-Boasts-More-Than-1-Billion-Smartphone-Users/1012984. Accessed 30 July 2018.

eMarketer. (2016, April 13). Smartphone user penetration in China, by age, 2014–2020 (% of mobile phone users). https://www.emarketer.com/Chart/Smartphone-User-Penetration-China-by-Age-2014-2020-of-mobile-phone-users/188732. Accessed 26 Mar 2019.

eMarketer. (2017). More than 95% of internet users in china use mobile devices to go online. https://www.emarketer.com/Article/More-than-95-of-Internet-Users-China-Use-Mobile-Devices-Go-Online/1015155. Accessed 14 Mar 2018.

Forbes. (2015, June 22). Environmental protests expose weakness in china's leadership. https://www.forbes.com/sites/forbesasia/2015/2006/2022/environmental-protests-expose-weakness-in-chinas-leadership/#4729b2017e63241. Accessed 31 May 2019.

Francesch-Huidobro, M., & Mai, Q. (2012). Climate advocacy coalitions in Guangdong, China. *Administration & Society, 44*(6S), 43–64.

Gao, R. (2013, December). Rise of environmental NGOs in China: Official ambivalence and contested messages. *Journal of Political Risk, 1*(8), http://www.jpolrisk.com/rise-of-environmental-ngos-in-china-official-ambivalence-and-contested-messages/. Accessed 25 July 2018.

Gevaert, R. (2012). A sustainable model for ICT capacity building in developing countries. In *26th Large Installation System Administration Conference (LISA 12), 2012* (pp. 123–134). Berkeley, C.A: USENIX Association.

Good, J. (2006). Chapter ten: Internet use and environmental attitudes – A social capital approach. In S. P. Depoe (Ed.), *The environmental communication yearbook* (Vol. 3, pp. 211–233). Abingdon: Routledge. https://doi.org/10.1207/s15567362ecy15560301_15567310.

Guo, C., & Saxton, G. D. (2014a). Tweeting social change: How social media are changing nonprofit advocacy. *Nonprofit and Voluntary Sector Quarterly, 43*(1), 57–79.

Guo, C., & Saxton, G. D. (2014b). Attention philanthropy: The good, the bad, and the strategy. *Nonprofit Quarterly.* https://nonprofitquarterly.org/2014/2008/2029/attention-philanthropy-the-good-the-bad-and-the-strategy/. Accessed 30 July 2018.

Hong, J. (2017, June). Social media in China: An unprecedented force for an unprecedented social change? *Telematics & Informatics, 34*(3), 691–693.

Howard, P. N., & Hussain, M. M. (2011). The role of digital media. *Journal of Democracy, 22*(3), 36–48.

Hsu, C., Chen, F.-Y., Horsley, J. P., & Stern, R. (2016, December 15). *The state of NGOs in China today.* Washington, DC: Brookings Institute. https://www.brookings.edu/blog/up-front/2016/2012/2015/the-state-of-ngos-in-china-today/. Accessed 31 May 2019.

Institute for Energy Research. (2015, May 20). China: World's largest energy consumer and greenhouse gas emitter. http://instituteforenergyresearch.org/analysis/china-worlds-largest-energy-consumer-and-greenhouse-gas-emitter/. Accessed 14 Mar 2018.

International Telecommunication Union (ITU). (2016). ICT Development Index 2016: IDI 2016 Rank. http://www.itu.int/net4/ITU-D/idi/2016/#idi2016countrycard-tab&CHN. Accessed 5 Aug 2017.

Jacobs, H. (2019, February 7). Alexandria Ocasio-Cortez included high-speed rail in her green new deal. After riding China's superfast bullet train that could go from New York to Chicago in 4.5 hours, it's clear how far behind the US really is. *Business Insider.* https://www.businessinsider.com/china-bullet-train-speed-map-photos-tour-2018-2015. Accessed 31 May 2019.

Jiang, S. (2017, August 17). The planned hydro plant will endanger the habit of rare peacocks in Yunnan Province. *China Youth.* http://big5.xinhuanet.com/gate/big5/news.xinhuanet.com/legal/2017-08/17/c_1121497433.htm Accessed 1 May 2019.

Kamrany, N. M., & Jiang, F. (2015, April 3). China's rise to global economic superpower. *HuffPost.* https://www.huffingtonpost.com/nake-m-kamrany/chinas-rise-to-global-eco_b_6544924.html. Accessed 19 June 2018.

Kaplan, A. M. (2012). If you love something, let it go mobile: Mobile marketing and mobile social media 4×4. *Business Horizons, 55*(2), 129–139.

Kepios. (2017, September 28). Digital in APAC in 2017 – IDUAI Special. https://kepios.com/special-posts/apac2017iduai. Accessed 30 July 2018.

Li, C., & Guo, J. (2015). Capacity-building development since the 1990s. *Asian Social Work and Policy Review, 9*, 79–93.

Li, Y., Koppenjan, J., & Homburg, V. (2017). Governing environmental conflicts: A comparative analysis of ten protests against industrial facilities in urban China.

Local Government Studies, 43(6), 992–1013. https://doi.org/10.1080/03003930.03002017.01375409.

Liu, C. (2014, November 6). China will limit pollution from steel and cement. *Scientific America.* https://www.scientificamerican.com/article/china-will-limit-pollution-from-steel-and-cement/. Accessed 5 Aug 2017.

McPhee, P., & Bare, J. (2001). Introduction. In C. J. De Vita & C. Fleming (Eds.), *Building capacity in nonprofit organizations* (pp. 1–5). Washington, DC: The Urban Institute.

Nah, S., & Saxton, G. D. (2013). Modeling the adoption and use of social media by nonprofit organizations. *New Media & Society, 15*(2), 294–313.

National Bureau of Statistics of China. (2018). *China Statistics Annual.* Beijing: National Bureau of Statistics of China. http://www.stats.gov.cn/tjsj/ndsj/2017/indexch.htm. Accessed 25 July 2018.

NonProfit Tech for Good. (2011, January 24). 22 ways nonprofits can use QR Codes for fundraising and awareness campaigns. http://www.nptechforgood.com/2011/01/24/22-ways-nonprofits-can-use-qr-codes-for-fundraising-and-awareness-campaigns/. Accessed 29 July 2018.

NonProfit Tech for Good. (2016, November 22). 5 nonprofit technology trends to watch in 2017. http://www.nptechforgood.com/2016/11/22/5-nonprofit-technology-trends-to-watch-in-2017/. Accessed 30 July 2018.

NPO Information Consultation Center. (2003). *NPO capacity building and the international experience.* Beijing: Huaxia Press.

Price, R., & Reuters. (2018, July 24). Facebook has created a subsidiary in China, even though its site is banned in the country. *Business Insider.* https://www.businessinsider.com/facebook-creates-china-subsidiary-ban-censorship-2018-2017. Accessed 27 July 2018.

Ryu, J. S. (2013). Mobile marketing communications in the retail environment: A comparison of QR code users and non-users. *International Journal of Mobile Marketing, 8*(2), 19–29.

Saleh, M., & Saifudin, M. (2017). Media and environmental non-governmental organizations (ENGOS) roles in environmental sustainability communication in Malaysia. *Discourse and Communication for Sustainable Education, 8*(1), 90–101.

Sangay, L. (2017, August 7). Tibet's fragile ecosystem is in danger. China must change its flawed environmental policy. *The Guardian.* https://www.theguardian.com/commentisfree/2017/aug/2007/tibets-fragile-ecosystem-is-in-danger-china-must-change-its-flawed-environmental-policy. Accessed 25 July 2018.

Schneider, S. (2014, October 15). Why NGOs and government organizations should use QR Codes. *uRQ.me.* https://uqr.me/blog/ngos-and-government-organizations-use-qr-codes/. Accessed 29 July 2018.

Sha, L. (2012, June 12). Environmental NGOs grow across China but still struggle for support. *Global Times*. http://www.globaltimes.cn/content/714330.shtml. Accessed 5 Aug 2017.

Shao, P., & Wang, Y. (2017). Social media in China: An unprecedented force for an unprecedented social change? *Telematics & Informatics, 34*, 694–704.

Tang, S.-Y., & Zhan, X. (2011, September 1). Political opportunities, resource constraints, and policy advocacy of environmental NGOS in China. Los Angeles: USC US-China Institute, University of Southern California. https://china.usc.edu/political-opportunities-resource-constraints-and-policy-advocacy-environmental-ngos-china. Accessed 31 May 2019.

Temple University. (2017). Digital media technologies, Minor. http://bulletin.temple.edu/undergraduate/intercollegial-programs/digital-media-technologies/. Accessed 19 June 2018.

The Economist. (2013, August 10). The East is grey. https://www.economist.com/briefing/2013/08/10/the-east-is-grey. Accessed 22 July 2018.

The Economist. (2015, March 12). Made in China? https://www.economist.com/leaders/2015/2003/2012/made-in-china. Accessed 31 May 2019.

UNESCO. (n.d.). Promotion of ICTs and media as tools for capacity building in peace and non-violence. *The United Nations*. https://en.unesco.org/cultureofpeace/main-areas-action/communicating-peace. Accessed 14 Mar 2018.

Wang, S. (2017, September 8). Why China can't get enough of QR codes. *CNN Money*. https://money.cnn.com/2017/09/08/technology/china-qr-codes/index.html. Accessed 2 June 2019.

WildAid. (2018, December 10). GOblue for a greener tomorrow. https://wildaid.org/goblue-for-a-greener-tomorrow/. Accessed 2 June 2019.

WildAid China. (2016, September 22). Grab a bike and GOBlue: New China campaign encourages low-carbon transportation. https://wildaid.org/grab-a-bike-and-goblue-new-china-campaign-encourages-low-carbon-transportation/. Accessed 2 June 2019.

Wilson, S. (2015, June 20). Mixed verdict on Chinese environmental public interest lawsuits. *The Diplomat*. https://thediplomat.com/2015/2007/mixed-verdict-on-chinese-environmental-public-interest-lawsuits/. Accessed 31 May 2019.

Wong, E. (2016, April 8). Clampdown in China restricts 7,000 foreign organizations. *The New York Times*. http://www.nytimes.com/2016/2004/2029/world/asia/china-foreign-ngo-law.html?_r=2010. Accessed 22 Nov 2016.

Yang, G. (2005, March). Environmental NGOs and institutional dynamics in China. *The China Quarterly, 181*, 46–66.

Zeitoun, T. (2003, February 19). *Capacity Building and ICTs: A Donor Agency Perspective*.... Paper presented at the WSIS PrepCom II. Geneva Roundtable 5.

The Multiple Airs: Pollution, Competing Digital Information Flows and Mobile App Design in China

Matteo Tarantino

Introduction

Air pollution is a key environmental issue in most industrialized countries. It also plays a complex and often subtle role in the production of localities or spaces with recognizable characteristics. Often, its influence on spatial production processes is mediated by *information* on air pollution, especially, since the early 2000s, when digitized public information about air pollution started circulating through the Internet. A lot of scholarship on this influence has focused on the efficacy of this information in driving the production of cleaner spaces, highlighting the presence in most contexts of issues of trust or disinterest. Comparatively less attention has been paid to the design of the vehicles of this digital information, on how and if they incorporate in their design such issues and on the solutions attempted to

M. Tarantino (✉)
Dipartimento di Scienze della Comunicazione e dello Spettacolo, Università Cattolica di Milano, Milano, Italy
e-mail: matteo.tarantino@unicatt.it

J. Díaz-Pont et al. (eds.), *The Local and the Digital in Environmental Communication*, Global Transformations in Media and Communication Research - A Palgrave and IAMCR Series,
https://doi.org/10.1007/978-3-030-37330-6_4

solve them. We argue that exploring the way environmental information vehicles are conceived and built enables a richer understanding of the relationship between pollution and space. This is because, as the Social Construction of Technology perspective has illustrated, unpacking the design processes of disclosure vehicles can provide insights into underlying social, cultural and technological processes, conflicts and tensions which may not be immediately visible to the analyst. This chapter will deal with one such case, a mobile app designed by a Chinese environmental NGO to circulate information about air quality in the People's Republic of China.

Theoretical Context: Space, Air Quality and Information

To locate the role of pollution in the construction of the 'local', it appears worthwhile to turn to the work of human (and phenomenological) geography and sociotechnical approaches towards a *relational* concept of space (Jones 2009). These approaches see space as emerging from the interplay amongst its material, symbolic and practice levels, drawing upon Lefebvre's idea of *production of space* (Lefebvre 1991). Recently, scholars have begun exploring the potential of the concept of 'translation' (Callon 1986) to describe the interrelationship between those levels (Tosoni and Tarantino 2013). Localities emerge, from this perspective, as the continuous interaction or 'translation' between the physical features of space, what social actors do with space and how space is represented. From this perspective, we can see how all three levels are directly impacted by air pollution. The most obvious is the material level. At the micro scale, air pollution 'can corrode metal and stone, discolor and dirty buildings outside and in, and in general make a neighborhood look shabby' (Ridker and Henning 1967). Moreover, it physically encumbers space, reducing visibility. At the macro physical level, air pollution impacts urban development and zoning via the pricing of real estate in both industrialized and emerging economies (Zheng et al. 2012; Zheng and Kahn 2008; Zheng et al. 2014), albeit determinants are debated (Chay and Greenstone 1998; Kim et al. 2003). At the level of practices, air pollution impacts on choices such as spending time outdoors and adopting and using air conditioners (Bresnahan et al. 1997) and on mobility patterns due, for example, to institutional restrictions on automotive transportation or individual changes of driving styles related to visibility (Knittel et al. 2016). At the representational level, air pollution can become a distinctive mark of spaces, to the point of transforming them into 'pollution landscapes' where 'environmental pollution occurs in such a way that the consequences of change become the dominant feature' (Broto et al. 2007). At the level of global media discourse, contemporary Beijing is a good example of a city whose representations are often dominated by air pollution (Xiang 2013), to the point of being declared the 'Air Pollution Capital in the World' (Watts 2005).

The degree to which of air quality affects spatial production is contingent on social actors perceiving, in some way or another, air pollution. If social actors are not aware of the pollution, many of the 'translations' between representations, practices and space fail to happen, and pollution becomes less powerful in contributing to the production of a locality. This is where information comes in. Below significant concentrations, most forms of air pollution are not immediately perceivable by the body. Moreover, air pollution can be normalized and discounted by social actors proportionally to the time spent in the polluted environments, in what is known as 'habituation' or 'adaptation' (Evans et al. 1981), albeit not all scholars agree on its extent (Menz 2011). Therefore, mediated communication about air quality becomes important to sensitize the population and reduce health risks. However, as mentioned, the effectiveness of mediated environmental information in actually changing people's behaviour is controversial at best, with studies indicating only a minority of subjects actually respond to pollution information with behavioural changes or an increase in environmental concern (Bickerstaff 2004). Research suggests that even subjects with respiratory issues tend to underestimate effects and do not adjust their daily practices (Wen et al. 2009). Most studies indicate bodily, direct perception of pollution as being the primary drive towards environmental concern and behavioural change and a 'distancing effect of information technologies' (Takács-Sánta 2007). When subjects perceive discrepancies between bodily experience and published information, trust in the latter tends to decrease substantially, especially when such information comes from official institutions such as local government agencies (Bickerstaff and Walker 1999). Further issues regard lack of access to and standardization of urban air quality data (Tarantino 2019, 2020), which can make cross-locality comparisons difficult. Lack of trust in air quality information has significant impacts on population, as some pollutants in the air can have severe health impacts even at concentrations hard to perceive by sight or smell, especially on weaker subjects such as the elderly and infants.

Weak trust or effectiveness have not impeded air quality information to become a mass commodity, especially after the widespread global adoption of the Internet. Digitized air quality information has become a massive information flow in many developed and developing countries (e.g. as of 2018, the global database World's Air Pollution reports AQIs from over 10,000 stations worldwide).[1] This has been fuelled by the development of legislative frameworks inspired by the principles of environmental justice and environmental transparency, which frame access to environmental information as a fundamental right. A milestone in this sense has been the *Convention on Access to Information, Public Participation in Decision-Making and Access to Justice in Environmental Matters*, promoted by the United Nations Economic Commission for Europe (UNECE) and subscribed in 1998 by 47 countries, which prescribes universal access to environmental information and in article 5(3) advocates explicitly for Internet means of distribution. As a result, in the past 20 years, we have witnessed an explosion in publicly available digital information about the environmental situation of many countries (van den Elshout et al. 2008).

Globally, the most widely available strand of this kind of information is air quality indexes or Air Quality Index (AQI)s. Air quality indexes are synthetic indicators of the degree of pollution in the air, based on the measurement of a set of pollutants. The typical range includes suspended particles (particle matters or PM), as well as other compounds and gases (mostly sulphur oxides, nitrogen oxides, ozone and carbon dioxides). These indexes are calculated through formulas which synthesize into single numbers the discrete measures of concentration of multiple pollutants produced by monitoring stations. The resulting number is mapped on a range whose breakpoints mark the different levels of air quality, usually ranging from 'excellent' to 'extremely bad'; the various levels are usually also colour-coded for more immediate understanding.

The formulas, breakpoints, choice of pollutants and even the colour-coding vary significantly between countries and even at the subnational levels, along with measurement techniques, location of the sensors and the timing of disclosure. Because of this lack of consensus AQIs have since the very beginning been contested measurements, around which considerable controversy exists (Longhurst 2005; Shooter and Brimblecombe 2008). Furthermore, AQIs have demonstrated issues in connecting with the very public they are designed to inform. Literature shows how abstract pollution

[1] http://www.aqicn.org/here, accessed on November 15, 2018.

communication, relying on numerical indexes, can often conflict with the physical experience of pollution (Bickerstaff and Walker 2001, 2003). When the two are at odds—for example, when people feel physical discomfort but AQI is low, or when AQI is high but little discomfort is felt—trust issues arise, and ulterior motives are attributed to AQI readings. In turn, the discourse of science is usually mobilized by institutions to ratify the primacy of AQIs and discredit bodily experiences as unreliable, situated and partial.

In each national context, we have therefore combinations of standardization, language and trust issues compounding on the capacity of air quality information to impact on space. In this chapter we are interested in exploring how these issues impact on the design of the vehicles of this kind of information. Our approach to the case study is twofold: first, the political/scientific scenario surrounding the design of the app will be reconstructed, to make the reader grasp the complexity of the tensions that the designers had to navigate. Subsequently, the evolution of the app will be illustrated and analysed.

Case Study

Methods and Goals

We selected a case study within a context characterized by (a) recognizable pollution issues; (b) high digital penetration; and (c) a controversial status of environmental information (i.e. the presence of an element of uncertainty with regard to this information). These coordinates imply three corollaries. First, the presence of an identifiable public need for environmental information. Second, that at least a part of this environmental information would be disseminated through digital means. Finally, that several competing solutions would be attempted, due to the controversy. Air quality communication in the People's Republic of China appeared to fulfil these requirements. Access to the design of the institutional disclosure platforms with the degree of detail needed by our approach has so far proven impossible in the country; thus the research focused on the specific design and production processes of a mobile app by a Chinese environmental NGO which throughout its history has enjoyed a positive relationship with the government.

Our methodology combines two main approaches. The first is a three-year ethnography (2014–2017) on the NGO where the app was designed. Aside from participant observation in the organization's activities and daily informal conversations (recorded in a research journal), we performed 31 semi-structured interviews with NGO members as well as 4 extensive interviews with the software developers of the app. Interviews focused on the history of development and particularly on the path of elimination of different possibilities towards the finished product. Moreover, two in-depth interviews with the US Embassy's officials, as well as an observational tour of the embassy premises, fed into the reconstruction of the air quality controversies which constitutes the background of our case and the content of the following section.

Our analysis was also informed by an approach to the software artefact, encompassing source code, interface and, for version 3.0, analysis of the various builds that succeeded towards the final version. By comparing the different versions of the artefacts, the analysis focused on charting the different approaches undertaken by the developers, including those which did not explicitly emerge from the interviews and could be fed back into subsequent interviews (see also Tarantino and Zimmermann 2017; Tarantino 2019).

Our analysis was completed by ten interviews with app users, five males and five females, all aged 25–35 and all members of Chinese e-NGOs. While this sample may appear limited and biased, it must be kept in mind that the research focused on exploring how existing tensions about air quality data would be incorporated at the design level and inscribed into the software artefact, rather than a systematic analysis of the audiences' reactions to the finished product. For this reason, our sample of users is chiefly composed of the same people the NGO used to informally test the various releases of the app.

Context: Establishing an AQI in China

Issues of environmental degradation in China, including the worsening of air quality, have been known for decades and have been documented outside of the PRC at least since the early 1980s (Smil 1984). International global attention has grown with the establishment of foreign professional communities in Chinese cities since the 1990s (Economy and Lieberthal 2007) and matured with the 2008 Beijing Olympics (Min and Zhen 2010; Tarantino and Carini 2010).

In China, especially in the urban areas, air quality information has come to be a daily presence in the life of hundreds of millions of citizens thanks to the penetration of smartphones, with a number of applications and websites offering information on air quality. As of November 2016, the

Baidu App Store, which by and large serves China alone, featured around 151 apps offering 'air pollution data' (*kongqi wuran shuju*)—to provide a comparison, the whole Google Play Store, which includes apps serving 145 countries, features around 240 comparable apps (including some also featured in the Chinese store).

By 1985, China had a particulate matter monitoring network which covered a total of 64 cities (Cao 1989), which expanded to 300 in 2001. Results were disclosed in the environmental statistical yearbooks published by the national environmental protection agency (which since 2009 became the Ministry of Environmental Protection). In 1996 the State Environmental Protection Agency (which would later become a ministry) created a national *Air Pollution Index* or API (*kongqi wuran zhishu*), a Chinese air quality index which lasted 14 years and two revisions (2000 and 2007), until in 2012 it was rebranded *Air Quality Index* (*kongqi zhiliang zhishu*). In February 1998, after much reluctance Beijing started to publish weekly air quality readings. By the Beijing Olympics of 2008, the Beijing Environmental Protection Bureau (henceforth EPB) was publishing daily API values on its website.

The transition between API and AQI was not smooth. API was based on normalized daily averages for sulphur and nitrogen oxides as well as PM10 (the largest of particulate matters), and hourly averages for carbon dioxide and ozone. It did not include measurements of finer particles, chiefly PM2.5. API readings were politically sensitive for two reasons. The first was the fear of social unrest, related to the impact on public health; this had been tempered by the notion that transparency would strengthen trust in institutions. The second was the strong government pressure with regard to better air quality. In 2006, in an effort of the central government to promote proactive policies towards air quality, performances of cities begun to be ranked not in terms of absolute pollutants concentrations, but in terms of the number of 'Blue Sky Days', that is, days with an API below 100. This requirement became particularly stringent around and during the 2008 Olympic Games held in Beijing. The local cadres' interest in having the highest possible number of 'Blue Sky Days' led to various critiques to existing air quality measurements in terms of sensors' positioning and reporting (Andrews 2008).

In 2007, the US Embassy of Beijing installed an air quality monitoring station. At the time, the embassy scientific attaché (who had trained as a nuclear physicist) reportedly grew dissatisfied with the discrepancy between her physical experience of the Beijing air and the API data being published every day by the Beijing Environmental Protection Bureaus. The published API on most days placed Beijing's air in the 'good' region of the scale. The initial set-up saw a small monitoring station hung outside the attaché's window; it was mostly analogue and 'local', as it offered air quality readings only to the people in its proximity, and the readings were not disclosed outside of the embassy. As technically US soil, the embassy's air quality monitoring system used the American United States Environmental Protection Agency's AQI and not the Chinese API, thus factoring in fine particles as well, particularly PM2.5. Through the EPA AQI, the embassy's machines ranked most of Beijing's days as 'unhealthy'.

The embassy decided this implied a health hazard for American citizens living in Beijing. This has significant implications: American embassies operate since 1990 under the 'No double standards' policy, which prescribes that information about risks run by US citizens abroad must be circulated to them (Hamilton 2011). Moreover, air quality hazards are ephemeral and may vary from one hour to the next; to fulfil the obligation, a continuous information system was therefore needed. Therefore, the subsequent step was to digitize this information and choose a platform to reach US citizens in Beijing on an hourly basis. The embassy chose the popular micro-blogging service Twitter, along with the embassy website. Starting from July 2008, the US Embassy started to tweet air quality information containing the AQI level and health recommendations (such as 'Unhealthy at 24/h exposure').

This chain of bureaucratic action/reactions brought about a political rift. At this point effectively two air quality rankings became available on the Internet for the city of Beijing, each with the full backing of an institution: the official API and the US Embassy AQI, with the latter consistently more alarming than the former. Understandably, the Beijing Environmental Protection Bureau and the Ministry of Foreign Affairs (MFA) were displeased by the discrepancy and tried unsuccessfully to pressure the embassy to limit this disclosure, citing disinformation and the creation of 'confusion' brought by the presence of multiple conflicting indexes. The Beijing EPB questioned the reliability of the US Embassy measurement. It questioned reliance on a single monitoring station, as opposed to the multiple stations used by the EPB, which granted a more reliable average; as well as its positioning, as the US monitoring station was set on the East Gate of the embassy, which overlooks the heavily trafficked Chang'an avenue, thus resulting in arguably higher-than-average smog readings (eventually the embassy station was moved on the South Gate, which overlooks an internal road and is therefore not as exposed to traffic, where it resides as of 2018). For its part, the Foreign Affairs office asked the embassy to adopt an outlet less public than social media for disclosure.[2]

As the rift intensified, a public campaign on Chinese social media platform Weibo, led by notable opinion leaders (Huang 2015; Chen et al. 2016), started to petition for a revision of the API in order to include information on PM2.5, the lack of which was a key cause of how low API readings appeared compared with the embassy ones. This revision was implemented in 2012,[3] with the new Air Quality Law, marking the progressive abandonment of API in favour of the new Chinese AQI (which would reach nationwide application only in 2016). The new Chinese AQI extends the covered pollutants to ozone, PM2.5 as well as hourly averages of SO_2 and NO_2. For most pollutants, and especially for PM2.5, the Chinese AQI is still more tolerant than the US AQI. This means it takes a higher concentration of pollutants to reach the same level of hazard.[4] Therefore, while the concentration numbers may be the same, there are considerable discrepancies in the AQI readings and therefore in the health recommendations offered to the public.

The issue was never really resolved. As of 2019, the US Embassy continues to publish its readings both on its website and on the Twitter account @beijingair. While Twitter is officially blocked in China since 2009, this has not impeded the spreading of alternative AQIs. On the one hand, the use of Virtual Private Networks (VPNs) to circumvent official

[2] US diplomatic cable (cable 09BEIJNG1945_a 7/10/2009) reports on the encounter between the US Embassy staff and the Chinese delegation consisting of representatives of the MFA and the Beijing EPB.

[3] The extent to which the new air quality standards have indeed been a direct response to the campaign is debatable. The extension to Ozone and PM2.5 was originally announced by the Ministry of Environmental Protection (since 2018, Ministry of Ecology and Environment) right before the Olympic Games of 2008, and four years appear to be a reasonable technical time for such a revision to be completed. However, it was certainly perceived as such, by campaigners, foreign media and scholars. See Xinhua (2008).

[4] Consider PM2.5, for example. To go from 'Good' air (AQI from 0 to 50) to 'Moderate' (AQI from 50 to 100), the Chinese Air Quality Index (the CNAQI) requires 24/hours average concentrations to be above 35 micrograms, whereas the US air quality information (USAQI) sets the breakpoint at 12 micrograms. To move from 'Moderate' to 'Unhealthy for Sensitive Groups', the ratio is 75/35.4, and from 'Unhealthy for Sensitive Groups' to 'Unhealthy' it's 115/55.4 However, for the subsequent two levels ('Very Unhealthy' and 'Hazardous'), for PM2.5, the difference between two AQIs is less than 1%. Significant discrepancies, averaging at 226%, regard also 24/h averages of SO_x and NO_x, and to a lesser extent PM1 particles (where the CNAQI is actually stricter of an average 3%).

blockings is quite widespread (and largely tolerated) in China. By using VPNs, bloggers and social media users could republish the embassy readings. Moreover, Twitter feeds are very easy to automatically mine through software scrapers. These scrapers, running on foreign servers, would then feed the scraped data to China, where again it could be embedded at will. This became particularly relevant after the smartphone explosion of the late 2000s, when a significant number of apps offering AQI data became available on the Chinese app market. The next section of this chapter will deal with one such case: the Blue Sky Map produced by the Institute for Public and Environmental Affairs, or IPE.

The Artefact: The Blue Map

IPE is an NGO founded in 2005 by environmental journalist-turned-activist Ma Jun. The organization works on information disclosure, and its core activity is collecting and releasing information about polluting factories. All of this information comes from official sources, mostly records of environmental violations filed by local Environmental Protection Bureaus. After the Open Government Information of 2007,[5] China has undergone a significant push towards the disclosure of certain kinds of official information, starting from environmental ones.[6] This reliance on official information (as opposed to user-generated or crowdsourced information) and the good reputation of the organization ensures it the amicable support of the Chinese central government, as well as of several provincial ones. It also ensures the organization a wide reach and a constant inflow of information, which translates into a proven ability to impact on transnational supply chains to drive environmental remediation in China. In short, reliance on official information is vital for the organization.

After relying almost exclusively on its website, in 2014 IPE launched its first mobile app, called *Wuran Ditu* ('Pollution Map'). Since Internet usage in China had shifted massively towards mobiles, the app was intended to target what multiple members of the NGO referred to as the 'average Joes' (in Chinese *Zhang San*), or lay public, whereas the website had come

[5] We are referring to the *Regulations of the People's Republic of China on Open Government Information (Zhengfu xinxi gongkai tiaoli)*.

[6] See in particular the 2008 *Measures on Open Environmental Information (for Trial Implementation) (Huanjing xinxi gongkai zanxing banfa)*. Many pieces of legislation have subsequently followed, expanding the available information.

to be used almost exclusively by the scientific and business communities. The cornerstone of the app, which set it apart from the NGO's website, was a map with hourly emission data from about 4000 large polluters designated by the government as 'Key State-Monitored Enterprises'. The idea was that this hourly data would be of greater interest to the lay user than data about environmental violations, because of being physically closer to users and to their immediate well-being. This information was coupled since the first release with localized air quality information. The idea was to 'give users the kind of information they need',[7] that is, information that would be pertinent to the users' immediate life-worlds. Air quality information looked like an 'obvious choice', the idea being that, in the words of one of the NGO members, '*in this way users will look at the app every morning and several times during the day to check for air quality*'.[8] In the design intentions, the attention grabbed by air quality information would lead users to the unique information contained in the app: the emission readings of the companies causing, at least in part, the pollution level expressed in the AQI number.

The second release of the app, rebranded 'Blue App' (*Weilan Ditu*, literally 'Blue Sky App') and published in 2015, further extended the AQI element of the previous one by allowing users to snap pictures which included an overlay with geo-localized information about air quality. In this way, pollution would acquire a further visual representation which users could share on social media to pressure the appropriate stakeholders (such as the Environmental Protection Bureaus, all of which are forced to have Weibo accounts to interact with the public)—with the added effect of popularizing the app.

When the app was released, the crowded Chinese app marketplace already featured a number of competitors (produced in China or abroad) offering multiple AQI readings, with the US Embassy one being consistently higher than the Beijing EPB averages. Because of its strategy to rely only on official government data, IPE continued to use only the official AQI readings. According to interviewees inside IPE, this may have had a more significant impact on the users' usage patterns than what the designers had originally imagined. For example, user Ms. Hu stated that she had '*stopped checking the app every morning* [after a few days of downloading

[7] M, 52. Founder. 2014 interview.
[8] F, 25, Communications Officer. 2015 interview.

the app] *to check air pollution, because I didn't trust the readings*'.[9] Like other respondents, Ms. Hu stated that this discrepancy was particularly felt on 'average' days when pollution was not massive. In days of massive pollution, AQI readings would be similar, and the bodily discomfort because of pollution would be reliable enough to prompt behavioural adjustments. On average pollution days, however, pollution would not be immediately visible or so harshly perceivable. Hence, as per our interviewees, AQI readings would be much more important for such choices as whether to bike to work, open one's windows, turn on air purifiers or jog in the park. Ms. Pang, one of the users who was also a member of another environmental NGO, stated that when in doubt '*you tend to trust the highest reading*'. This appeared to be true, according to our respondents, irrespective of the mentioned technical shortcomings of the US Embassy reading (chiefly it being a single-point measurement in a city of 16,000 square kilometres). This seems to indicate that the official AQI information incorporated by IPE in the app might have lost some of its traction due to the availability of competing readings on the Chinese Internet. In other words, users did not open the app in the morning or during the day to check AQIs, because other apps already provided them with information they felt as more needed: the one with the more critical reading. This appeared to be confirmed by the app analytics data, which showed a limited number of daily active users (about 5% of the total download base as of 2015). The NGO went back to the drawing board for a second overhaul of the app with the objective of increasing the number of active users by increasing the amount of 'relevant information' provided to them. In doing so, it maintained its choice to use only the official AQI readings.

The third main release of the app was launched in May 2016 and included several new features. Concerning AQI, the core innovation was to integrate it with weather information and provide recommendations about daily practices related to atmospheric conditions. This advice was expressed through icons in the lower part of the main screen and included such topics as 'wearing a mask', 'exercising outdoor', 'washing your car', 'wearing sunscreen' and so on. These tips are calculated by an algorithm provided by an external company, but on thresholds defined by IPE. This solution was paralleled by other ongoing attempts at 'repackaging' AQI information in other forms. The Chinese app Moji Weather (*Moji Tianqi*, 墨迹天气), which offers a system of recommendations through a very

[9] F, 26, 2016 Interview.

similar interface, was quoted during the design phase as a direct inspiration. A less direct inspiration, quoted by some members of the NGO but not directly by the design team, was the popular French app Airpocalypse. At the time of writing, the app offers daily humorous daily quips and recommendations related to air quality, such as 'Highway to the Danger Zone' for an AQI of 325, or the recommendation: 'masks are out of style' for an AQI of 55. Airpocalypse also states to offer readings 'from the US consulate where possible' (the practice of monitoring air quality at US embassies and consulate has expanded to, as of 2018, 33 other cities).[10]

However, compared to those apps, the Blue App and IPE are, in principle, not primarily concerned with air quality, but with identifying and pressuring polluters. Air quality is incorporated as a pull factor for users, with the pollution source data as the intended payload of their engagement. The third and fourth revisions of the app make this connection between different kinds of information even more explicit: when opening the lifestyle tips, users are presented with information about the environmental transparency performance of relative brands, compiled by IPE itself (e.g. by opening the tip about car washing, users would see performances of car manufacturers). Moreover, they include full weather and wind information and forecasts, so as to extend the timeframe of relevance of air quality information.

Implications

Our research dialogues with one of the core questions of this volume: how environmental communication is impacted by the intertwining of the local and the digital. Addressing this question appears particularly urgent in the case of environmental information, because of the tension between the strongly *local and embodied* character of the object of environmental communication (particularly related to air quality) and the distance entailed by its abstraction and institutionalization (further complicated, in our case, by multiple institutionalizations). This tension in turns has significant impact on trust, and ultimately on the capacity of environmental information to be an effective agent for change towards more sustainable living spaces.

[10] See https://airnow.gov/index.cfm?action=airnow.global_summary. Consulted on November 30, 2018.

We will explore this question by discussing two core implications of our case study. The first regards the attempt to maximize the ability of the Blue App to contribute to the production of a cleaner space by sidestepping the trust issues. Using the terminology of this volume, the ***digital*** attempting to act on the ***local***. The Blue App is designed to entail functions dealing with each of the impacts of air quality over urban space production in the sense delineated in the first paragraph. At the level of materiality, the app features a function specifically designed to document physical changes in polluted spaces, through its overlaid photo function.[11] At the level of representations, the app features maps allowing users to identify 'polluted cities' (using a city-level average AQI) and, by zooming in a specific city, to segment space into variously polluted neighbourhoods according to localized AQI readings. And the latest releases made explicit the connection with the practice level, with its adoption of lifestyle recommendations. From this perspective, the app can be read as a tool to foster the production of a less environmentally hazardous space, primarily by increasing environmental awareness: users can gain an understanding of the air quality of their locality and pressure, directly or indirectly, for change, also modifying their own set of practices.

Digital information is in this sense the indispensable, primary vector of this transformation: the app is designed to exploit the widespread incorporation of portable technologies into daily routines, for example, by sending the user a daily notification with the AQI condition.

At the same time, the digital is also the vector of competing readings of AQIs, which hinder the efficacy of the app, a situation which emerges from the specificities of the local—in this case, a particular environmental-techno-juridical-political arrangement happening in the city of Beijing in the second half of the 2000s. In Beijing in particular, the traditional trust issues of AQIs appeared, at the time of our observation, to be intensified by the availability of multiple institutionalized air pollution indexes due to the conflict between the stiffness of two institutional arrangements (the US Embassy's 'No double standards' rule and the Chinese national sovereignty over the definition of pollution standards), the relative porosity of the Chinese Internet (irrespective of relatively tight formal regulation) and a considerable demand for pollution information due to the country's environmental crisis. Our case study shows a situation in which 'local air

[11] Notice how also unpolluted landscapes are produced by the app, such as when 'Blue Sky Days' are celebrated with pictures with low AQIs by users.

quality' is a contested, undecided measure for the citizens themselves, where the usual conflict between 'body' and 'numbers' is supplemented by another conflict between two different 'official' numbers. While our project did not include enough research on audiences to draw reliable conclusions, our small sample of users seems to point to the fact that, in doubt, even knowledgeable users tend to trust the most negative information. For sure, this notion was acknowledged by the designers.

This leads to the second implication: we can observe in our case study forms of deinstitutionalization of air quality information. The trust crisis was not an abstract issue for the designers of the app, as it effectively hindered usage of the app itself. The design response has been to accompany the abstraction of numeric values (to which IPE is bound) with an approach to the daily life of users. The addition to meteorological conditions information is a part of this strategy: air quality is packaged alongside less contentious information such as the weather. Yet here we are still in the institutional realm. A second, more complex step appears to be the repackaging of AQIs into lifestyle recommendations: this design choice shows the potential to circumvent the trust issues. Specific lifestyle tips, as opposed to the broad health recommendations which are correlated with AQI levels, tend to be less prone to direct comparisons and are calculated at the discretion of IPE itself, not of public authorities. Moreover, their formulas are not visible to the end user and are thus less contestable than air quality standards. At the same time their specificity has the potential to create interruptions in the daily routines of users (such as disincentives to go outside, wash one's car, etc.) and thus weaken the normalization of pollution perception. In this choice, we can see air quality information being (at least in part) deinstitutionalized, appropriated and transformed by civil society to circumvent the trust issues. The digital allows in this sense relatively ample room to experiment without much economic cost. On the other hand, balancing the institutional with the deinstitutionalized is particularly delicate in countries such as China. Indeed, in our case, the AQI number remains the foremost element of the app interface, testifying to the enduring primacy of the institutional information.

Finally, the solution adopted from the designers can also be read as an attempt to reinvolve the body of the user, by situating communication close to the daily life of the user. We have seen how the body appears to hold a primacy in determining adaptations to air quality issues; yet, in most air quality communication bodies are abstracted as a vessel of health, and the communication of risk is reduced to indications about possible health hazards. In the case of air quality issues whose impacts on health are mostly incremental and long-term, this may not have been a winning strategy to drive the production of cleaner spaces. More research is needed to test whether referencing pollution impacts on users' daily life—the things users do every day in space with their bodies—can mend the 'distancing effect' of mediated air quality communication.

Conclusions

This chapter described how the design of a software artefact intended to foster environmental changes in space came to embed cultural, political and scientific conflicts, and has highlighted the corresponding response strategies. We have seen how long-standing issues of trust in air quality information (deriving from a core conflict between bodily experience and techno-scientific discourse) came to be exacerbated by a conflict between the ease of circulation of digital information and the rigidity of institutional arrangements. We have seen how to retain its power to act on the local, the digital is deinstitutionalized and packaged in forms closer to the daily life of users; if proven impactful by more systematic future research on audiences, this may trace a potential path for more effective environmental communication in other countries. Finally, our research suggests that the study of environmental risk communication in the era of pervasive computing should include accounts of software design and production to better understand the localized impacts.

Acknowledgements This chapter has been developed under the CHIPOMAP research project, funded by the Swiss National Science Foundation grant #153291 and directed by Dr. Basile Zimmermann. I am grateful to the Institute, Ma Jun and all of IPE's present and past staff for the continuing support, to the editors and to the two anonymous reviewers for the useful feedback on the first draft.

References

Andrews, S. Q. (2008). Seeing through the smog: Understanding the limits of Chinese air pollution reporting. *China Environment Series, 10*, 5–32.

Bickerstaff, K. (2004). Risk perception research: Socio-cultural perspectives on the public experience of air pollution. *Environment International, 30*, 827–840.

Bickerstaff, K., & Walker, G. (1999). Clearing the smog? Public responses to air-quality information. *Local Environment, 4*, 279–294.

Bickerstaff, K., & Walker, G. (2001). Public understandings of air pollution: The 'localisation' of environmental risk. *Global Environmental Change-Human and Policy Dimensions, 11*, 133–145.

Bickerstaff, K., & Walker, G. (2003). The place(s) of matter: Matter out of place – Public understandings of air pollution. *Progress in Human Geography, 27*, 45–67.

Bresnahan, B. W., Dickie, M., & Gerking, S. (1997). Averting behavior and urban air pollution. *Land Economics, 73*, 340–357.

Broto, V. C., Tabbush, P., Burningham, K., Elghali, L., & Edwards, D. (2007). Coal ash and risk: Four social interpretations of a pollution landscape. *Landscape Research, 32*, 481–497.

Callon, M. (1986). Some elements of a sociology of translation: Domestication of the scallops and the fishermen of St. Brieuc Bay. In J. Law (Ed.), *Power, action, and belief: A new sociology of knowledge* (pp. 196–223). New York: Routledge.

Cao, H. (1989). Air pollution and its effects on plants in China. *Journal of Applied Ecology, 26*, 763–773.

Chay, K. Y., & Greenstone, M. (1998). *Does air quality matter? Evidence from the housing market.* Washington, DC: National Bureau of Economic Research.

Chen, W., Tang, H., & Zhao, H. (2016). Urban air quality evaluations under two versions of the national ambient air quality standards of China. *Atmospheric Pollution Research, 7*, 49–57.

Economy, E., & Lieberthal, K. (2007). Scorched earth: Will environmental risks in China overwhelm its opportunities? *Harvard Business Review, 85*, 88.

Evans, G. W., Jacobs, S. V., & Frager, N. B. (1981). Air pollution and human behavior. *Journal of Social Issues, 37*, 95–125.

Hamilton, D. (2011). The transformation of consular affairs: The United States experience. In J. Melissen & A. M. Fernandez (Eds.), *Consular Affairs and Diplomacy.* Leiden, Brill.

Huang, G. (2015). PM2.5 opened a door to public participation addressing environmental challenges in China. *Environmental Pollution, 197*, 313–315.

Jones, M. (2009). Phase space: Geography, relational thinking, and beyond. *Progress in Human Geography, 33*, 487–506.

Kim, C. W., Phipps, T. T., & Anselin, L. (2003). Measuring the benefits of air quality improvement: A spatial hedonic approach. *Journal of Environmental Economics and Management, 45*, 24–39.

Knittel, C. R., Miller, D. L., & Sanders, N. J. (2016). Caution, drivers! Children present: Traffic, pollution, and infant health. *Review of Economics and Statistics, 98*, 350–366.

Lefebvre, H. (1991). *The production of space.* Oxford: Blackwell.

Longhurst, J. (2005). 1 to 100: Creating an air quality index in Pittsburgh. *Environmental Monitoring and Assessment, 106*, 27–42.

Menz, T. (2011). Do people habituate to air pollution? Evidence from international life satisfaction data. *Ecological Economics, 71*, 211–219.

Min, W., & Zhen, X. (2010). Mirroring the Olympic games – The Beijing 2008 Olympic games in the American media. *The International Journal of the History of Sport, 27*, 1794–1808.

Ridker, R. G., & Henning, J. A. (1967). The determinants of residential property values with special reference to air pollution. *The Review of Economics and Statistics, 49*, 246–257.

Shooter, D., & Brimblecombe, P. (2008). Air quality indexing. *International Journal of Environment and Pollution, 36*, 305–323.

Smil, V. (1984). *The bad earth. Environmental degradation in China*. London: Zed Press.

Takács-Sánta, A. (2007). Barriers to environmental concern. *Human Ecology Review, 14*, 26–38.

Tarantino, M. (2019). In the air tonight: The struggles of communicating about urban environmental quality. In Z. Kraijina & D. Stevenson (Eds.), *The Routledge urban media and communication companion*. London: Routledge.

Tarantino, M. (2020). Navigating a datascape: Challenges in automating environmental data disclosure in China. *Journal of Environmental Planning and Management, 63*(1), 67–86.

Tarantino, M., & Carini, S. (2010). The good, the fake and the cyborg: The broadcast and coverage of Beijing 2008 Olympics in Italy. *The International Journal of the History of Sport, 27*, 1717–1738.

Tarantino, M. & Zimmermann, B. (2017). Database Green: Software, Environmentalism and Data Flows in China. *The China Quarterly, 229*, 205–217.

Tosoni, S., & Tarantino, M. (2013). Media territories and urban conflict: Exploring symbolic tactics and audience activities in the conflict over Paolo Sarpi, Milan. *International Communication Gazette, 75*, 573–594.

Van Den Elshout, S., Léger, K., & Nussio, F. J. E. I. (2008). Comparing urban air quality in Europe in real time: A review of existing air quality indices and the proposal of a common alternative. *Environment International, 34*, 720–726.

Watts, J. (2005). China: The air pollution capital of the world. *The Lancet, 366*, 1761–1762.

Wen, X. J., Balluz, L., & Mokdad, A. (2009). Association between media alerts of air quality index and change of outdoor activity among adult asthma in six states, BRFSS, 2005. *Journal of Community Health, 34*, 40–46.

Xiang, D. (2013). China's image on international English language social media. *The Journal of International Communication, 19*, 252–271.

Xinhua. (2008). Ozone, particle pollution to be included in monitoring. *The China Daily*, 2008/08/05.

Zheng, S., & Kahn, M. E. (2008). Land and residential property markets in a booming economy: New evidence from Beijing. *Journal of Urban Economics, 63*, 743–757.

Zheng, S., Wu, J., Kahn, M. E., & Deng, Y. (2012). The nascent market for "green" real estate in Beijing. *European Economic Review, 56*, 974–984.

Zheng, S. Q., Cao, J., Kahn, M. E., & Sun, C. (2014). Real estate valuation and cross-boundary air pollution externalities: Evidence from Chinese cities. *Journal of Real Estate Finance and Economics, 48*, 398–414.

SECTION II

Social Media, Environmental Protest and Community-Building

Localizing Environmental Conflicts: Facebook Groups as Intertextual Sites for Local Protest Voices

Anders Horsbøl

Introduction

Green transition has given rise to new conflicts relating to construction of energy devices such as wind turbines, hydroelectric power plants, solar heating systems, and biogas plants that have been established not least to reduce carbon emissions and mitigate climate change. These devices have been opposed on the grounds of concerns about the quality of life for local residents, economic effects, the impact on the landscape, biological diversity, and others. This has led to heated conflicts, particularly among local inhabitants, the media, and municipalities, and also with national authorities, NGOs, and academic experts involved on both sides. The conflicts illustrate that, even if green transition is agreed to be

A. Horsbøl (✉)
Department of Communication and Psychology, Aalborg University, Aalborg, Denmark
e-mail: horsboel@hum.aau.dk

J. Díaz-Pont et al. (eds.), *The Local and the Digital in Environmental Communication*, Global Transformations in Media and Communication Research - A Palgrave and IAMCR Series,
https://doi.org/10.1007/978-3-030-37330-6_5

a desirable goal, its realization is not just an instrumental question of implementation, but involves negotiation between different concerns.

The notion of locality or place is central in these conflicts, in a triple sense. First, the conflicts thematically evolve around the possible construction of a renewable energy device at a specific site. The conflicts are about something at a particular locality, not about renewable energy, wind turbines, or solar energy in general. Second, place is often invoked in a more emphatic sense as possessing a special value for a particular group of people, because of qualities of the natural environment, economic development, or personal attachment. Finally, the conflicts are often played out in a local public arena, that is, in the local press or on social network sites, with participants from a specific area as a driving force.

This chapter will address the issue by examining voices against the construction of wind turbines at specific localities in Denmark as they play out on protest sites on Facebook. The Danish experience is of wider interest, since Denmark is among the world frontrunners when it comes to the use of wind energy for electricity supply, both land based and offshore. In 2017, wind turbines produced 43.7% of the electricity supply in Denmark, a share that has increased steadily from 19.4% in 2009.[1] Protests against the construction of wind turbines have occurred since the 1990s, but have increased over the last decade, as more of the turbines have been deployed. The Landsforeningen Naboer til Kæmpevindmøller (LNK) (National Association of Neighbors of Gigantic Wind Turbines) was founded in 2009 and lists on its website more than local 200 protest groups, most of them from the Jutland peninsula in the west of the country. As a result of the protests, projected wind turbine projects have been postponed or abandoned by several municipalities. Organizations countering the protests exist as well, for instance the Dansk Vindmølleforening (Danish Wind Turbine Owners' Association), but these seem to be less vocal in the local debates. Organizations such as Danmarks Naturfredningsforening (Danish Society for Nature Conservation) have generally welcomed wind turbines, but at the same time raised concerns about their placement at particular locations.

Instead of simply dismissing the protest voices as NIMBYism ('not in my back yard'), it is arguably more productive to examine the arguments and concerns actually put forward (Devine-Wright 2009). This chapter will do so using a discourse analytic approach, focusing on the ways in

[1] Source: www.windpower.org

which the Facebook groups ascribe meaning to environmental interventions at particular locations.

By understanding the protest Facebook sites as issue publics (Bruns and Highfield 2016) and applying a network approach to the public sphere, this chapter pays particular attention to the import and appropriation of discursive material from other public spaces. Operationalized via discourse studies notions of intertextuality and recontextualization, the analysis examines how the Facebook sites bring in a variety of other texts and voices and how the appropriation of these in the digital space helps make sense of controversies over wind turbines.

Theoretical Background

Although climate change and green transition have received considerable attention within the field of environmental communication in recent years (see, for instance, Eide et al. 2010; Stevenson and Dryzek 2014; Hansen and Machin 2015; Painter et al. 2016), conflicts over the erection of environmental energy devices at specific localities have hitherto only been studied to a limited degree. The few studies that do exist consider conflicts over wind turbines (Rudolph 2014; Boon and Reimer 2015; Horsbøl 2019), a hydroelectric power plant (Carvalho et al. 2016), and a tidal power plant (Ko et al. 2011). The studies demonstrate that green conflicts are not limited to a specific geographic area, but can be found at several sites around the globe. However, further studies are needed to explore these conflicts in more depth.

From the perspective of environmental communication, the conflicts are interesting for at least two reasons. First, at a practical level, they seem to pose a significant challenge to green transition, at least given that "[t]his war over 'green' concepts will occur more and more around the world in the twenty-first century" (Ko et al. 2011, p. 15). Second, at the level of principle, the conflicts point to the interpretability of notions such as 'sustainability,' 'environmental,' and 'green transition.' The conflicts provide examples of negotiations and struggles over the meaning of these more general notions, in particular over what a fair and just green transition should look like. The practical and principal levels are connected, however, since the struggles illustrate that, even if green transition is deemed to be a desirable goal, realizing it will need to take into account different understandings of what constitute environmental values and how the different values are to be weighed against one another. To realize green transition is therefore not just an instrumental question of implementation, but involves negotiations relating to what it means to care about the environment and how the different meanings can be integrated. Understanding the specific conflicts over the erection of wind turbines, as in the present study, may thus prepare us for dealing constructively with other struggles to come in the process of green transition.

The conflicts over environmental energy devices play out in different public spaces. As one of these spaces, Facebook represents a media format, often referred to as 'social media' or 'social network sites,' which allows participants to sidestep the gatekeeper function of journalism and communicate directly with a wider public. As such, Facebook offers new opportunities also for environmental communication, not least for individuals or groups which may not have easy access to the news media. However, it is worth stressing that Facebook should not be seen in isolation, but as an increasingly integrated part of the public sphere and its flows of discourse in a 'hybrid' media system (Chadwick 2013).

In the public debates on wind turbines in Denmark, a large number of Facebook groups have been formed by protesters at different localities. They confront the erection of wind turbines in a specific area, stressing the consequences for a specific community. As such, they form instances of *issue publics* (Dahlgreen 2009) or *sphericules* (Bruns and Highfield 2016), which have a narrower thematic focus and exist on a shorter time span than the traditional news media. However, in tackling their specific issues, they also draw on sources and bring in voices from the wider public sphere. They can therefore be seen in the framework of public sphere conceptualizations that stress connections, networks, and flows between different arenas, rather than events within one public arena (Habermas 1992; Mansbridge et al. 2012; Chadwick et al. 2016). Within this framework of a *networked public sphere*, it is important to ask how and to what extent different publics are connected, how themes and voices move from one arena to another, and what these discursive exports and imports mean for the specific public arenas. These questions guide the present study.

Tentatively, the protest Facebook groups may also be described as *counter publics* (Fraser 1990; Kaiser and Puschmann 2017). According to the latter researchers, a counter public "opposes the hegemonic view and is excluded by the mainstream and/or excludes itself from the mainstream to regroup" (Kaiser and Puschmann 2017, p. 373). Whether this is the case for the Facebook groups in the present study remains to be investigated. However, as this study focuses on the communication in selected

Facebook groups, the question of an exclusion by 'the mainstream' lies outside the study. Moreover, the delimitation of 'the hegemonic view' or 'the mainstream' may not be entirely clear. Nevertheless, the notion of a counter public can be used in an *emic* sense to investigate how the protest Facebook groups construe their relation to other publics and whether the idea of countering a mainstream or even a hegemonic position plays a significant role in the postings on the site.

The present study attempts to combine the network approach to the public sphere with a *discourse* approach. At a more general level, a discourse approach to environmental issues studies the ways in which meaning is ascribed to environmental interventions, such as the erection of wind turbines, and how different concerns are articulated and negotiated. It is furthermore deemed productive here to combine the network perspective on the public sphere with discourse studies notions of *intertextuality* and *recontextualization* (Linell and Sarangi 1998; Chouliaraki and Fairclough 1999; Horsbøl 2005; Fairclough 2010; van Leeuwen 2009; Vásquez 2015), inspired by the works of Bakhtin (1986), concerning intertextuality, and Bernstein (1990), concerning recontextualization. The concept of intertextuality directs our attention to the ways in which texts draw upon and are interwoven with other texts, thereby pointing to the historicity of the individual text. This presence of other texts can take many forms (see, for instance, Genette 1992), but will be understood in the present study in a more restricted and operative sense as *manifest intertextuality* (Fairclough 2003). That is, intertextuality refers to the manifest presence of other texts, for instance in the form of a quotation, explicit reference, or link. Correspondingly, the concept of recontextualization draws our attention to the ways in which discourse travels and is modified in new situations. Recontextualization thus implies that meaning is modified or renegotiated when discursive material is transferred from one context to another. The point is not the intertextual connection as such, but the way in which it is conducted and how it gives rise to new meaning. In that sense, intertextuality and recontextualization can be used as parallel notions to study how the selected Facebook sites draw upon material from wider public arenas, for instance from national mass media, neighbourhood municipalities, or other protest sites, and how this import of discursive material contributes to making sense of the local issues at stake. In this way, the study explores how the local debate extends beyond the local and examines how discursive import from other localities adds to the local controversy.

Data and Analytical Procedure

The chapter presents case studies of two Facebook groups organized by protesters against the erection of wind turbines at two different sites. Based on an overview of several Danish Facebook groups opposing wind turbines, the two cases have been chosen in accordance with the principle of maximum breadth in qualitative, explorative studies on three parameters: placement of wind turbines, activity level, and forms of intertextuality. Concerning placement, the one Facebook group, Stop Vesterhav Syd Havvindmøller[2] (SVSH) ('stop offshore wind turbines in the southern North Sea'), opposes offshore wind turbines along the western coast of the Jutland peninsula, whereas the second group, Nej til kæmpevindmøller i Aabenraa Kommune[3] (NKAK) ('no to gigantic wind turbines in the council district of Aabenraa'), opposes land-based turbines in the southeastern part of Jutland. With regard to activity, the SVSH group has a high level, with sometimes several posts a day, whereas the NKAK group has a lower level, with posts on average once or twice a week. Third, with regard to intertextuality, the two groups differ considerably in their way of bringing in events, issues, and voices from other publics, as is explored in the analysis. SVSH was established on October 10, 2015, has 807 followers, and appears to be run by an organization opposed to wind turbines at Vesterhav Syd. NKAK was established on January 29, 2015, has 470 followers, and provides no further organizational information apart from that they are "citizens in the council district of Aabenraa who do not want the municipal plans for giant wind turbines to be carried out."[4] Owing to the difference in the frequency of postings, the following analysis is based on posts on the SVSH page from January to June 2018 and on the NKAK page from January 2017 to December 2017. All posts in the period have been read through and, following the identification of main activities at the two sites, an archive of screen dumps representing the different activities in each group has been collected.

The analytical strategy has been to commence in an inductive fashion by asking the basic questions: What happens on these two Facebook sites? What kinds of discursive action take place? In answering these questions, the analysis proceeds by conducting a short overview of the main activities

[2] https://da-dk.facebook.com/stopvesterhavsyd/

[3] https://da-dk.facebook.com/Nej-til-k%C3%A6mpevindm%C3%B8ller-i-Aabenraa-Kommune-457394237744080/

[4] Assessed June 28, 2018.

in the two arenas. Thereafter, the identified activity types are analysed in more depth. In part of the analysis, the notion of topos (Bornscheuer 1976; Zompetti 2006; Wengeler 2013) is used to identify the commonly employed premises on which opposition to the erection of wind turbines in the local area is based. A topos thus refers not to what is claimed, but to the perspective from which something is argued, identifiable mainly via lexis and semantic relations. The analysis of activities pays particular attention to the intertextual practices of linking, referring to, and commenting on other texts, since this appears to be prevalent in the two Facebook groups.

Analysis

Overview of the Activities in the Facebook Groups

The activities in the two Facebook groups can be divided into at least three main types: (a) self-made campaign material, (b) calls for action, whether online or offline, and (c) links to and comments on other texts, mostly from publicly accessible websites. The latter activity is clearly predominant in both Facebook groups.

Compared to other types of Facebook activity, the *absence of discussion* in the two protest groups is striking. The groups do not offer long threads with diverging opinions on an issue. The commenting field rarely exceeds two or three comments, and there is little disagreement between the comments. Thus, the two Facebook groups are not used as an arena for exchanging points of view among the participants. This does not mean that the participants are not engaged in public discussions on the issue, or that the Facebook group is not connected to a debate on this issue, but the discussions take place outside the Facebook page, as will be shown later in the chapter.

Main Activities: (a) Campaign Material

A smaller amount of the material consists of videos or images produced by the group to promote messages opposing the erection of wind turbines in the area. In this respect, the Facebook site functions as a platform for the distribution of campaign material, which could just as well be found on organization websites. In fact, this is the case with the SVSH group, which also owns a website (http://www.stopvesterhavsyd.dk/), where some of the same material can be found.

Fig. 1 Screen shots from the video "Naturens rige" ("The realm of nature")

An example of this promotional material is the video "Naturens rige" ("The realm of nature") on the SVSH Facebook site, which presents some of the key arguments at the SVSH site in a condensed form.[5] Without sound, the video shifts between the use of three semiotic forms: motion pictures or still pictures from the local area, motion pictures from the area with animated wind turbines attached, and text items with a short message in a big white font on a black background. Screenshots of the three forms are shown in Fig. 1.

At the beginning of the video, the term 'the realm of nature' is shown in quotation marks, referring to the way in which the municipality speaks about the local coastal area. This is followed by a handful of still images of the coastal landscape, showing a sunlit beach and the sea in wide shots, with just a few humans taking up little space against the background of sand, sea, and sky. These images all seem to confirm the characterization of the area as a realm of nature.

Next follows a section on tourism, stating in numbers the alleged impact on the area in turnover and jobs. The text messages are supplemented with still pictures of people on the beaches and in pedestrian streets.

In the next section, a text item states that "gigantic wind turbines are now planned 4–8 km from the coast," followed visually by animations of wind turbines on the horizon, all with circulating blades. Textually, it is claimed that a wind turbine park may provide 35 temporary jobs, while 300 jobs will be lost in tourism, and that "the whole of Denmark will forever lose a unique piece of nature." Thus, the affected community is extended beyond the local area. To illustrate this point again, the concluding images show the landscape first without wind turbines and then with

[5] https://www.facebook.com/stopvesterhavsyd/videos/vb.415682078624664/460984687427736/?type=2&theater. The video lasts two minutes and has had more than 6000 views within 2¼ years.

animated turbines added. The video closes with a text item saying that "the local support is not there, and has never been there," thereby questioning the democratic legitimacy of the wind turbine project.

All in all, the video coarticulates three topoi in the sense of premises for evaluating the wind turbine plans. First, the consequences for tourism and the local economy are highlighted textually. Second, a topos of landscape is foregrounded in one text screen by reference to the potential loss of "a unique piece of nature" and expressed in the contrasting images of the coast with and without wind turbines. Third, a topos of democracy is evoked by reference to the alleged lack of local support. This topos is present only in one text, but at a prominent place at the very end. In the flow of the video, the topos of landscape establishes the premise of the topos of tourism or local economy, realized in the reference to estimated lost jobs in local tourism, and those two topoi further strengthen the weight of the topos of democracy, in the sense that the lack of democratic legitimacy is yet more problematic when the consequences of a non-democratic decision are dire. The evidence associated with the three topoi is presented through numbers and animations of what the future landscape will look like. Neither of these is substantiated by further warrants, but the animated images with wind turbines in particular endow the representation with a sense of an immediate insight into the future. This can be understood as an example of *evidentia*, not in a verbal form concerning past events, as in the classical tradition of rhetoric, but by visual means anticipating the future.

By contrast, the NKAK site contains no videos produced by the group for the particular local case. Instead, a few images can be found of posters opposing the erection of wind turbines in the area. The images appear to be photos from public (protest) meetings and draw on a topos of nature similar to that seen in the SVSH video. Thus, whereas the SVSH site offers local campaign material that is purely digital, the NKAK site represents existing material from outside the digital realm.

Main Activities: (b) Calls for Action

The two Facebook groups have several posts that call for action from the reader. An example is the following post on the NKAK site:

> (1) Den kommunale valgkamp starter snart og man kan møde TvSyd og komme med forslag til valgkampsemner. Vi synes det er en god mulighed for at sætte vindmøller på som valgtema. Aabenraa den 9/9 kl 10.30-13.30 i Føtex. (NKAK, September 8, 2017)

["The municipal election campaign is about to begin, and there will be a chance to meet TV-Syd [a regional television channel] and put forward suggestions for election campaign issues. We think this is a good opportunity to put wind turbines on the election agenda. Aabenraa, September 9, 10.30–13.30 in Føtex [a supermarket]."]

The post calls for action in relation to an upcoming municipal election by attempting to influence the agenda of a regional television channel. It illustrates how the two Facebook groups often call for action beyond the digital realm and how connections are made between the digital (the post), the physical co-presence (in the supermarket), and the news media (the television channel). While this post promotes agenda-setting in the news media, similar posts include suggestions to contact politicians, show up at council meetings, or sign petitions.

In addition to these specific suggestions, several calls for actions at a more general level can be found, as the following text from an SVSH post shows:

(2) Det er JER som følgere, der kan gøre en forskel – ikke denne fb-side alene. I skal DELE relevante og oplysende opslag her fra SVSH-siden, helst til beslutningstagere og landsdækkende medier – og skrive Læserbreve – ellers er der ikke nogen, der opdager og forstår vores ønske om at flytte møllerne ud til 20–30 km fra kysten. Deltag også i Vattenfalls info-møde skærtorsdag. (SVSH, March 22, 2018)

["YOU are the ones who can make a difference – not this FB page alone. You need to SHARE relevant and informative posts from the SVSH page, preferably to decision makers and national media – and write letters to the editor – otherwise no one will hear about and understand our desire to move the wind turbines 20–30 km away from the coast. Also go and take part in Vattenfall's [the provider of the wind turbines] information meeting on Maundy Thursday."]

Using a considerably more prescriptive tone, this post instructs the readers—who are presumed to agree with the aims of SVSH—not to be passive consumers of the information on the Facebook page, but to participate by distributing relevant information and by approaching the media and decision-makers. The post gives clear instructions as to who, how, and why the actions should be undertaken, formulated mainly in general terms without reference to particular persons or events. The last sentence, however, adds a specific event to the list, a local meeting with the wind turbine company Vattenfall, which is to be in charge of establishing and running the planned wind turbine park. As with the quoted NKAK post, the call for action covers digital action often related to social network sites (sharing), action associated with the traditional news media (writing letters to the editor), and action associated with co-presence (showing up at a meeting). Whereas these fields of action can be distinguished theoretically, it is significant that they are often interwoven at the two Facebook sites. In addition, the SVSH post reaches beyond the local in the sense that national media are mentioned as a target group. This widened scope can be seen as parallel to the extension of the affected community to the whole of Denmark in the SVSH video analysed above.

Main Activities (c): Linking

The most common type of activity in both Facebook groups is to link to other internet texts. The linking is mostly done by the use of a hyperlink, which is partly visible on the post and can be accessed by clicking. For the most part, the other text is not only linked to but also commented upon in the posting in a way that briefly relates it to the issue of the Facebook group. The typical posting in the two Facebook groups thus consists of a link to another text on the internet and a comment that semantically links the issue of the other text to the issue of wind turbine erection in the local area. The following section presents examples of the types of texts typically linked to, the ways in which this linking is done semantically, and the resulting contribution to the local issue.

In both Facebook groups, articles in the *local or regional newspapers*, both news and views, are often linked. In most cases, the reference text can be clicked on, but in the NKAK group, photos of newspaper articles, that is, without hyperlinks, are occasionally posted. By following the link to the local paper, a lively discussion in the comments field on the newspaper site can often be observed. As mentioned earlier, there is little discussion in the two Facebook groups, but the links to the local paper connect the Facebook groups to local debate. Thus, the issue public of the Facebook group seems, to a large extent, to be dependent on the local newspaper for representing the local debate on the wind turbine issue.

The links to the newspaper arena can be found both with and without comments. The comments often briefly evaluate the text linked to, as in example (3), or present a reframing of it, as in example (4). In (4), the first sentence simply repeats the title of the newspaper article, while the second phrase in brackets adds the new framing that the political division mentioned in the title results from one faction refusing to see the problem.

(3) Dette rigtig gode læsebrev var i JV igår! MÅ MEGET GERNE DELES! (NKAK, October 27, 2017)

["This very good letter to the editor was in Jydske Vestkysten [regional newspaper] yesterday. YOU ARE VERY WELCOME TO SHARE IT!"]

(4) Artikel: Møller deler Venstrefolkene (Der er nogen der ikke vil se problemet!) (SVSH, March 21, 2018)

["Article: Wind turbines divide the people in Venstre [political party]. (Someone refuses to see the problem!)"]

It is also quite common to link to *other protest sites*. This goes particularly for the NKAK group, which links heavily to posts (including videos) from the above-mentioned National Association of Neighbors of Gigantic Wind Turbines (LNK). As part of its operation, LNK distributes, that is to say, delocalizes, information on local events beyond their municipal borders. This sort of information can then be taken up, that is, re-localized, by other local protest groups, who may reflect their own experiences in stories from neighbouring or related municipalities. The national LNK may thus function as a hub that enables a flow of information and recontextualization between different local issue publics.

The NKAK group frequently shares posts from the LNK site consisting of links to related local stories. For example, a shared post from the LNK page consists of a link to a news story in the national online media site altinget.dk about discussions between a representative from the Danish wind turbines industry and mayors from municipalities that have turned down plans to erect wind turbines. Whereas the original article on altinget.dk referred to disagreements between the two sides on reasons for rejecting wind turbines, LNK highlights a quote by one of the mayors:

(5) "Hvis ét eneste menneske bliver sygt af at have en vindmølle for tæt på, så er det aldrig nogensinde noget, jeg vil have i Viborg Kommune," siger Torsten Nielsen og støttes af Henrik Frandsen. (LNK, March 26, 2017)

["'If a single human being is going to fall ill by having a wind turbine too close, then that is never ever something I would want in the council district of Viborg', says Torsten Nielsen [mayor of Viborg council], and he is supported by Henrik Frandsen [mayor of Tønder council]"]

The LNK thus focuses attention on a particular person's reasoning. This refocusing is elaborated by the NKAK group when it shares the LNK post, by adding that "some mayors speak up against the wind turbine industry" (NKAK, March 26, 2017). This two-step linking allows the original story to be reinterpreted as being about speaking up against the wind turbine industry. Moreover, in the local context, the reference to "some mayors" who speak up may of course be read as a contrast with the local mayor. Thus, the linking not only communicates information on similar controversies in neighbouring councils but also appropriates the information for rhetorical use in the local controversy.

A third way of linking consists of the *enrolment of other parties.* Posts in the SVSH group cite the local trade association, the national Dansk Ornitologisk Forening (Danish Ornithological Association), and the national Danish Society for Nature Conservation as allies. The latter is particularly interesting, since it is among the largest NGOs in Denmark, with more than 130,000 members, and since its rationale resonates with the topos of landscape put forward by the SVSH group. An example is a post on January 26, 2018, in which SVSH links to a letter to the editor in the local paper from representatives of the local branch of the Danish Society for Nature Conservation. The comment summarizes the content of the letter as "support from the Danish Society for Nature Conservation."

Reference to *scientific knowledge* is a fourth way of linking, which is similar to the enrolment of other parties, but differs in that authoritative knowledge instead of specific interests is allegedly referred to. For example, NKAK links to a post by LNK, which again shares a news article from the online section of the national television channel TV2. The news article reports on a recent study by the Danish Cancer Society on the relationship between noise and cancer of the intestine. The news article, however, is focused on traffic noise and does not mention noise from wind turbines. The semantic link to wind turbines is made by LNK by adding the following comment, quoting a report from the Danish Health Authority.

> (6) Rapport udført for Sundhedsstyrelsen: "Støjgene er den væsentligste effekt af støj fra vindmøller. Støjgenen fra vindmøller er større end for vejtrafikstøj ved samme støjniveau." (LNK, July 6, 2017)
>
> ["Report produced for the Health Authority: 'Noise nuisance is the main effect of noise from wind turbines. The noise nuisance from the wind turbines is greater than from road traffic at the same level of noise.'"]

The Health Authority report quoted in the comment, however, is from 2011, and is not referred to in the news article linked to by LNK. Thus, it is the comment by LNK alone that connects the shared news article to wind turbine noise and at the same time brings two authoritative sources on health knowledge together (the Health Authority and the Cancer Society). This association is then imported to the NKAK group without further comment. This example may illustrate both the sometimes lengthy and complex intertextual chains in the recontextualization process from scientific reports to Facebook issue publics and the opaque intertwining of different scientific sources in that process.

Finally, a fifth way of linking may be termed *issue enlargement*. Here, other issues are made relevant for and coupled to the wind turbine issue. This form of linking mainly takes place in the SVSH group, concerning both specific and general issues. A specific issue is, for example, a debate on a planned extension of the electricity supply in the western part of Denmark, which would involve a series of electricity poles with high voltage bestriding long stretches of the landscape. This plan has given rise to some controversy and local opposition, and the SVSH consistently links to news stories on the issue with comments that explain the need for new electricity poles with the planned wind turbine parks opposed by the group. For instance, in linking to a story in the local newspaper *Jyske Vestkysten*, SVSH calls on a national politician involved in the electricity pole issue to also bring up the planned wind turbine parks, since "it is those that have caused the cable run through Jutland" (SVSH, June 3, 2018).

An example of a more general issue is tourism. In this case, the SVSH posts often comment on the news stories on tourism with the claim—in line with the argumentation in the video analysed above—that the very condition for tourism is an uncluttered coastal landscape, and that it is precisely this landscape that is threatened by the planned wind turbines. For instance, a new government plan for increasing tourism is linked to and met with the comment that "acceptance of new coastal wind turbines and an increase in coastal tourism – is very self-contradictory!" (SVSH, June 8, 2018). In both the specific and general cases, the SVSH group not only technically links to these other issues, but also semantically construes them as causally related to the wind turbine issue. Thus, the local controversy on wind turbines is interwoven with other issues and made relevant for other debates.

Conclusion

This chapter has studied two Facebook groups opposing the erection of wind turbines in two specific areas in Denmark. By applying a network perspective on the public sphere in combination with a discourse studies focus on intertextuality and recontextualization, the Facebook groups have been studied as issue publics, in which meaning is ascribed to the planned local erection of wind turbines, not least via discursive import from other sites. The analysis shows that linking practices play a major role in the two groups and that the groups thereby appropriate material from other public arenas in ways that are instrumental to their own local aims. Although the two Facebook groups are concerned about a local issue, they draw heavily on other (internet) sources outside the local area. Moreover, the linking practice goes beyond the technical linking enabled by the Facebook functionality, since the external material is appropriated in systematic ways by semantic linking, such as by enrolment of other parties or enlargement of the issue.

The two Facebook pages also differ in their linking practices. The NKAK group often links to the National Association of Neighbors of Gigantic Wind Turbines, not least with stories from other localities, which are then re-localized at the NKAK site. The SVSH group, on the other hand, is far more active in connecting the local issue of wind turbine erection to other public issues, such as tourism or electricity supply.

Apart from the linking practices, the two Facebook groups also contain a limited amount of more traditional campaign material. The discursive construction in this material is generally in line with the meaning making that emerges from the linking practices. The sites thus display a discursive consistency across different activities.

The two Facebook sites abound with calls for action, both within and outside the digital realm. These calls concern action on social networks sites, in relation to traditional news media, and for physical co-presence. The calls also include national players and national media. Thus, although the Facebook sites are local and digital, they also extend beyond those bounds.

Discussion—in the sense of exchanges between different points of view—is scarce in the two Facebook groups. More discussion takes place in the local press, to which the Facebook groups heavily link. This is an indication of the importance of the local press as a provider of a debate on crucial environmental decisions.

Whereas the Facebook groups clearly illustrate the network character of the modern (digital) public sphere, it is less obvious whether they can be understood as counter publics. This characterization may be validated by the fact that many of the postings oppose what is viewed as a mainstream position. On the other hand, the efforts, in particular by the SVSH group, to enrol allies among mainstream players, to side with the council ideals of the local area, and to appropriate issues and enlarge the thematic field controvert the characterization of the groups as counter publics. At least for the SVSH group, opposition to the local wind turbine plans takes place also by aligning with the mainstream. This opens up new questions for public debate, such as the following: Do the enrolled parties agree to be enrolled, and do others agree that the appropriated issues are related to the wind turbine issue? Whether these directions can prove fruitful to the environmental controversy remains to be investigated.

Summing up on the local and the digital in environmental struggles, the present study leads to the following conclusions. The two Facebook sites do not represent a local space in the simple sense of a space marked off from the national or from other localities. Of course, the Facebook sites circulate information on local events, but, more interestingly, the local plays out often as a *localization* of events, issues, and debates which originate elsewhere. Thus, as locally based issue publics, the Facebook groups function to a considerable extent by making non-local phenomena locally relevant. Apart from the technical linking, this is done by a complex of discursive practices such as reframing, issue enlargement, and enrolment of other parties. In this respect, the local is not a limited space, but a specific form of connectivity.

In a similar way, the digital does not appear as a realm separated from other (analogous) fields of activity, but rather as a way of *digitizing* different forms of activity by representing them in the digital space. This includes activity that is to a high degree non-digital (such as protest meetings), activity that is simultaneously both (the press), and activity that is mainly digital (other websites). These types of activity are not kept separate in the digital issue publics studied, but are interwoven and coarticulated.

Finally, it follows that the meeting of the digital and the local is not the connection of two separate realms, but rather *the intertwining of two types of connectivity*. The present study has only begun exploring what this implies for the understanding of environmental communication.

References

Bakhtin, M. M. (1986). *Speech genres and other late essays*. Austin: University of Texas Press.

Bernstein, B. (1990). *The structuring of pedagogic discourse*. London: Routledge.

Boon, T. E., & Reimer, T. (2015). Blæst om demokratiet. *Momentum, 4*, 28–31.

Bornscheuer, L. (1976). *Topic. zur Struktur der gesellschaftlichen Einbildungskraft*. Frankfurt a.M.: Suhrkamp.

Bruns, A., & Highfield, T. (2016). Is Habermas on Twitter? Social media and the public sphere. In A. Bruns, G. Enli, E. Skogerbø, A. O. Larsson, & C. Christensen (Eds.), *The Routledge companion to social media and politics* (pp. 56–73). New York/Oxford: Routledge.

Carvalho, A., Pinto-Coelho, Z., & Seixas, E. (2016). Listening to the public – Enacting power: Citizen access, standing and influence in public participation discourses. *Journal of Environmental Policy and Planning, 12*, 563–576. https://doi.org/10.1080/1523908X.2016.1149772.

Chadwick, A. (2013). *The hybrid media system: Politics and power*. Oxford: Oxford University Press.

Chadwick, A., Dennis, J., & Smith, A. P. (2016). Politics in the age of hybrid media. Power, systems and media logics. In A. Bruns, G. Enli, E. Skogerbø, A. O. Larsson, & C. Christensen (Eds.), *The Routledge companion to social media and politics* (pp. 56–73). New York/Oxford: Routledge.

Chouliaraki, L., & Fairclough, N. (1999). *Discourse in late modernity. Rethinking critical discourse analysis*. Edinburgh: Edinburgh University Press.

Dahlgreen, P. (2009). *Media and political engagement: Citizens, communication, and democracy*. Cambridge: Cambridge University Press.

Devine-Wright, P. (2009). Rethinking NIMBYism: The role of place attachment and place identity in explaining place-protective action. *Journal of Community & Applied Social Psychology, 19*(6), 426–441.

Eide, E., Kunelius, R., & Kumpu, V. (Eds.). (2010). *Global climate – Local journalisms: A translational study of how the media makes sense of climate summits*. Bochum/Freiburg: Projektverlag.

Fairclough, N. (2003). *Analysing discourse. Textual analysis for social research*. London: Routledge.

Fairclough, N. (2010). *Critical discourse analysis: The critical study of language* (2nd ed.). London: Longman.

Fraser, N. (1990). Rethinking the public sphere: A contribution to the critique of actually existing democracy. *Social Text, 25–26*, 56–80.

Genette, G. (1992 [1979]). *The architext: An introduction*. Berkeley: University of California Press.

Habermas, J. (1992). Further reflections on the public sphere. In C. Calhoun (Ed.), *Habermas and the public sphere*. Cambridge: MIT Press.

Hansen, A., & Machin, D. (Eds.). (2015). *Visual environmental communication*. London: Routledge.
Horsbøl, A. (2005). The GM beer as discursive object taking place: Towards an analysis of the materiality of discourse. In M. Asplund Carlsson, A. Løvland, & G. Malmgren (Eds.), *Multimodality: Text, culture and use*. Kristiansand: Høyskoleforlaget.
Horsbøl, A. (2019). Green conflicts in environmental discourse. A topos based integrative analysis of critical voices. *Critical Discourse Studies*. https://doi.org/10.1080/17405904.2019.1617757.
Kaiser, J., & Puschmann, C. (2017). Alliance of antagonism: Counterpublics and polarization in online climate change communication. *Communication and the Public, 2*(4), 371–387.
Ko, Y., Schubert, D. K., & Hester, R. T. (2011). A conflict of greens: Green development versus habitat preservation – The case of Incheon, South Korea. *Environment: Science and Policy for Sustainable Development, 53*(3), 3–17.
Linell, P., & Sarangi, S. (Eds.). (1998). Discourse across professional boundaries. Special Issue, *Text, 18*(2), 143–318.
Mansbridge, J., Bohman, J., Chambers, S., Christiano, T., Fung, A., Parkinson, J., Thompson, D. F., & Warren, M. E. (2012). A systemic approach to deliberative democracy. In J. Parkinson & J. Mansbridge (Eds.), *Deliberative systems – Deliberative democracy at the large scale* (pp. 1–26). Cambridge, UK: Cambridge University Press.
Painter, J., et al. (2016). *Something old, something new: Digital media and the coverage of climate change*. Oxford: Reuters Institute for the Study of Journalism.
Rudolph, D. (2014). The resurgent conflict between offshore wind farms and tourism: Underlying storylines. *Scottish Geographical Journal, 130*(3), 168–187.
Stevenson, H., & Dryzek, J. S. (2014). *Democratizing global climate governance*. Cambridge, UK: Cambridge University Press.
van Leeuwen, T. (2009). Discourse as the recontextualization of social practice: A guide. In R. Wodak & M. Meyer (Eds.), *Methods of critical discourse analysis* (pp. 144–161). London: Sage.
Vasquez, C. (2015). Intertextuality and interdiscursivity in online consumer reviews. In R. H. Jones, A. Chik, & C. A. Hafner (Eds.), *Discourse and digital practices. Doing digital analysis in the digital age*. London: Routledge.
Wengeler, M. (2013). Historische Diskurssemantik als Analyse von Argumentationstopoi. In D. Busse & W. Teubert (Eds.), *Linguistische Diskursanalyse: neue Perspektive* (pp. 189–215). Wiesbaden: Springer VS.
Zompetti, J. P. (2006). The value of topoi. *Argumentation, 20*(1), 15–28.

Surviving the Legacy Media System: The Place of Local Digital Activism in the Chiloé Red Tide Crisis

Claudio Broitman and René Jara

Introduction

In this chapter, we analyse how an argument that linked a red tide phenomenon in southern Chile (2016) to the local salmon industry was spread by activists throughout Facebook and digital media. This link between the red tide and the salmon industry was reproduced far beyond the initial controversy and accusations, in a country that is highly connected to online networks, with 13 million Facebook accounts (2018). An unusually strong red tide phenomenon was the origin of a serious socio-environmental crisis at the beginning of 2016 in southern Chile. Strong scientific evidence proves that the red tide is a harmful algal bloom (HAB) due to natural processes (like temperature, wind or biomass changes). There is insufficient evidence to affirm that nutrients caused by human activity promote the growth of harmful species and the red tide (Davidson et al. 2014) and yet, this was the core argument of the activists. These

C. Broitman (✉) • R. Jara
School of Journalism, University of Santiago de Chile, Santiago, Chile
e-mail: claudio.broitman@usach.cl; rene.jara@usach.cl

J. Díaz-Pont et al. (eds.), *The Local and the Digital in Environmental Communication*, Global Transformations in Media and Communication Research - A Palgrave and IAMCR Series,
https://doi.org/10.1007/978-3-030-37330-6_6

acknowledged, anthropogenic factors may increase nutrients (like nitrogen and phosphorus) in the water. Not only coastal aquaculture (with a strong local presence), but also organic wastes, fertilizers or atmospheric inputs tend to elevate nutrients concentration in coastal waters. This is known as eutrophication. However, the occurrence of HABs does not necessarily imply eutrophication (Davidson et al. 2014). Red tide's toxic microorganisms can be dangerous both for animals and for humans that consume infected animals. These microorganisms can cause many problems ranging from food intoxication to death by cardiopulmonary arrest. As a consequence, the commercialization of seafood during the red tide is impossible. This prohibition can generate social collateral effects, especially in territories depending on natural resources coming from the sea.

During the 2016 red tide phenomenon, legacy media (including press and broadcasting service) widely covered the controversy. While some stakeholders held that the phenomenon was due to natural causes, some locals and activists maintained that it was related to anthropogenic factors. On the one hand, local communities assumed that the red tide was related to the salmon industry, well established in the region. On the other hand, alongside concerns related to the salmon industry, legacy media published scientific arguments that denied the connection between the salmon industry and the red tide. Our case explores Chiloé Island, an isolated territory with a rich cultural, historical and natural life in southern Chile, where salmon and fishing industry have located their farms for 50 years.

Social movements have increased in Chile in recent years, particularly in relation to social-environmental conflicts. Various cases include the local opposition against the mining exploitation of Pascua Lama in the Huasco Valley (Urkidi 2010) and the most emblematic case, HidroAysén, a mega dam project in the Chilean Patagonia that incited a strong movement across the country (Reyes and Rodríguez 2015) and a controversy surrounding scientific and lay knowledge production (Broitman and Kreimer 2018). During the 2016 red tide controversy, the major claim from local activists' was that 9000 tons of aquaculture salmon died because of the red tide—and were dumped into the sea, thus generating uncertainties of a possible link between that dump and the red tide itself. These claims of connection between salmon industry practice, a mass fish kill—and its local impacts on the livelihoods and health of local Chilean communities—were dispersed through the Facebook network. This local controversy's survival beyond legacy media in the digital networks that connect local activists to local communities in Chile and throughout the world are examined here.

Literature Review and Theoretical Framework

Scholars have studied salmon industry environmental conflicts, especially in the main producing countries Norway, Chile, Canada, the United States and the United Kingdom (Scotland), from different perspectives. Some research has focused on the way the industry comes into conflict with other relevant stakeholders and the responses of the administration and public policies (Noakes et al. 2003; Singleton 2009; Daniels and Mather 2017). For example, Harrison and Loring (2014) point to the institutional failure characteristic of these types of conflicts. Institutions often understood conflicts without considering complex aspects such as people's cultural identity and previously established relations of power or trust. The role of the Norwegian press in these types of salmon controversies and conflict (Osmundsena and Olsen 2017) has also been examined, with a focus on the way in which stakeholders establish their arguments and circulate them in the public debate. They found how media debates, built by discourse coalitions, don't introduce new arguments over time, rather repeat the same rhetorical concepts and storylines. The Chilean red tide example supports this scholarship that seeks to better understand the complexity of the salmon industry, beginning with local activists and their role in the environmental conflict that often accompanies this industry. Findings point to stakeholder strategies to disperse their interpretations about the red tide crisis through social and digital media and in doing so establish, influence and, ideally, define the parameters of public debate around the controversy.

These strategies to spread a specific interpretation of the crisis constituted a favourable context for the emergence of intense social movement's activity (Della Porta and Diani 2011), particularly in social media. The possibility of these communication tools to challenge established corridors of power have been widely disseminated (Castells 2007). Much of the substantive discussion in this work has shown that activity in social media is important to enabling connective action (Bennett and Segerberg 2012). Its use was observed with particular interest in the case of the Arab Spring, where technologies were observed as a trigger for the organization of protests (Mansour 2012; Youmans and York 2012; Lim 2013; Wolfsfeld et al. 2013). In the Chilean case, the relation between activists and social media has been recently studied too, particularly in the protest movements of students. For example, Valenzuela et al. (2012, 2014) found that a systematic use of Facebook enhances the chances to participate in a protest activity among high school students.

According to Cammaerts (2015), there are eight key roles in which activism employs social media. These include internal organization, recruitment and links; mobilizing for and coordinating direct action; dissemination of the activist frames independently of the main currents; discussion, debate, deliberation and decision-making; attacking ideological enemies; surveilling the surveillants and the preservation of protest artefacts (pp. 4–5). In this chapter, we only observed three of these functions: (a) the dissemination of activist frames; (b) the discussion/debate/deliberation and decision-making functions, but, above all, (c) the surveillance of the surveillers. Our findings conclude that it is in this way that the local and activists' core argument—that the red tide phenomenon was related to anthropogenic factors—retained its legitimacy and currency in ongoing public sphere debate. Thus, local activists occasionally permeate the legacy media agenda (Sapiezynska 2014), working as media that connects and monitors social conflict.

During the period of the Chilean political transition in the 1990s, the general trend was the invisibility of this type of conflict by legacy media (Bresnahan 2003). The current trend marks a significant change in the Chilean local context. Digital media provides activists with the tools to challenge the invisibility of protests and its communication or not, by legacy media (Lester and Hutchins 2012). In this way, digital media seem to have played a very important role (Gronemeyer 2013), particularly in the context of concentrated media ownership (Geoffroy and Sunkel 2001; Becerra and Mastrini 2017). In some cases, it has supported a renewal of the conflict and its key argument. In our case, the dynamics of media consumption seem consequently to be influenced by "news in social media" (Godoy and Gronemeyer 2012, p. 51). The media content of the conflict produced in this context interacts strongly with audiences, to some extent bypassing and challenging the established power of legacy media. It is sometimes the only way in which citizens are informed of distant and complex events. This consumption of news seems to be especially relevant in social-environmental conflicts, because the information in those contexts is unclear and the definitions and details of the litigation itself are under dispute too. In these contexts, social media can be a more accessible source of communication than legacy media.

Ironically enough, alternative media disappeared with the return to democracy in Chile. Well-known opposition journals and magazines—*Diario La Época, Análisis, Apsi*—stopped publishing after the return of democracy. These types of publications that emerged in the heat of the struggle against the dictatorship were shut down, because they didn't have any budget to continue (Bresnahan 2003). In recent years though, many digital media (González-Bustamante and Soto 2016), as well as other media projects such as community radio (Ramírez Cáceres 2010) and the recent addition of digital TV (Sáez-Baera 2013), have reopened to challenge the legacy media space, making inroads in the Chilean media market. It is no coincidence that these media have incorporated socio-environmental conflicts into their agenda, with the will to represent local communities and the voiceless. This role is reminiscent of alternative media that emerged to challenge the Chilean dictatorship and is now 'digital' and 'social'.

Methodology

Our methodology analyses the activity of digital activists that used social networks during the 2016 red tide controversy. Our data consists of three Facebook *fan pages*[1]—'Chiloé está privao', Centro de Estudios Sociales de Chiloé (CESCH) and Defendamos Chiloé—and the posts published between 2016 and 2018 ($n = 2937$). We collected this data with a special application called 'Netvizz', a tool that extracts data from the Facebook platform. The selection criteria had three components: we selected the *fan pages* operated by activists strongly rooted to the territory who have actively participated in the debate on social networks.

We divided our research design into two key steps. Firstly, we focused on the Facebook *fan pages* between 2016 and 2018. Here, we especially relied upon the engagement index.[2] The engagement index is a three-variable combination of number of shares, likes, reactions and comments for each post published. Secondly, we analysed the contents of those posts from a qualitative perspective. Using the collected data, we identified five recurrent topics: (a) the polluting salmon industry/against the salmon industry; (b) the link between the salmon industry and the red tide; (c)

[1] A Facebook fan page is a site dedicated to a person or a one cause. A fan page differs from a personal profile because it is visible for everyone.

[2] For more detail, see: http://developers.facebook.com/docs/reference/login/

the link between the salmon dumping and the red tide; (d) the red tide and (e) the social movement. Both these methods allowed us to observe the different strategies among the activists, especially in terms of activity surrounding the controversy over an extended period.

The Facebook Sites

The three Facebook accounts correspond to three organizations that are quite diverse and further outlined below. All three have a political stance contrary to the prevailing economic development model and against institutions that support growth as the key aim in societies. In this sense, these three Facebook communities are linked to the critical discussions regarding extractivism—a current that is also present in much of the socio-environmental conflicts of the continent. Extractivism is a type of "natural resource removal" that we can see in mining, agriculture, forestry and oil extraction (Gudynas 2014, p. 12). Latin America and Chile are mainly oriented to the export of natural resources, which consolidates them as a supplier region for developed countries (Wallerstein 1998).

'*Centro de Estudios Sociales*'[3] *(CESCH)* is the local anthropologist's organization opposed to the extractivist development model. For example, they have been keen on communicating one of the Chiloé's biggest problem: the negative effects of deforestation and the destruction of peat land on freshwater underground storage capacity. They have a headquarter in Castro[4] and are self-financed.

During the period under examination, 295 posts were published in the 'Centro de estudios sociales de Chiloé' (CESCH) Facebook *fan page;* 158 corresponded to shared links, 75 to photographs, 39 to videos, 13 to status and 10 to events. Approximately half of these messages were posted in 2016 (155), 125 in 2017 and 15 in 2018. In terms of metrics, the post that reached the highest level of engagement reached a peak of 3377 likes in May 2016 (Fig. 1).

'Chiloé ta'privao'[5] is one of the most politicized groups. They are more committed in offline activities, participating in the mobilization of fishermen, reporting their news and progress in their negotiations with the

[3] In English it means, Center of Social Studies from Chiloé.

[4] Chiloé's capital, an island located on southern Chile.

[5] A game of words with two different senses: it means on the one hand that *Chiloé is private* and on the other hand that *Chiloé don't have something*.

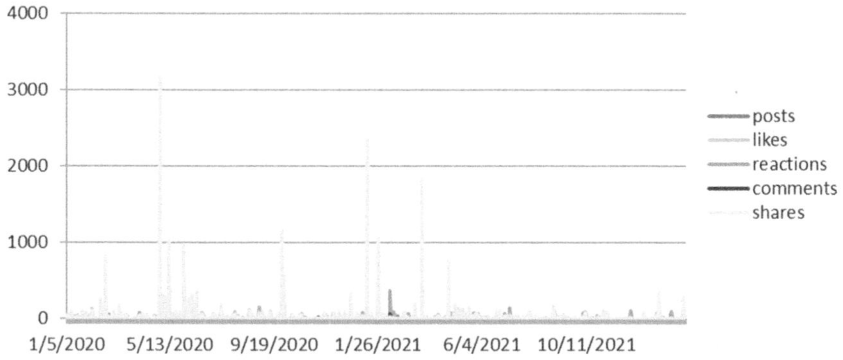

Fig. 1 'Centro de estudios sociales de Chiloé' data *fan page*. (Source: Fan page engagement based on data downloaded with Netvizz)

government. Their goal is to achieve concrete measures regarding the island social demand.

During the period under examination, 384 posts were published in the Facebook 'Chiloé ta'privao' *fan page*. The posts are divided as follows: 62 correspond to videos, 62 to status, 179 to photos, 67 to links and 14 to events. 344 were posted in 2016, 38 in 2017 and 2 in 2018. The top 20 with the highest level of engagement include videos, photos and links. The message that had the greatest impact had an engagement of 3733 likes. Eighteen of these 20 messages were posted in May and June of 2016 (Fig. 2).

'*Defendamos Chiloé*'[6] is a project that openly revealed its physical address in the area. They are invited to forums and the local press uses it as an information source. The 'Defendamos Chiloé' *fan page* is the only one that preceded our study. It exists before the red tide crisis and it remained for years after that.

In this *fan page* there are 2268 posts published in the time interval studied. The posts are divided as follows: 1762 correspond to photos, 205 to videos, 199 to links, 86 to status and 16 to events. A total of 1154 messages were posted in 2016, 770 in 2017 and 344 in 2018. Photographs are among the top 20 publications with the highest level of engagement; from those photographs, 15 were published during May 2016. Within the post they hold a higher level of engagement with the highest reaching 29105 score (Fig. 3).

[6] In English it means, "let's defend Chiloé".

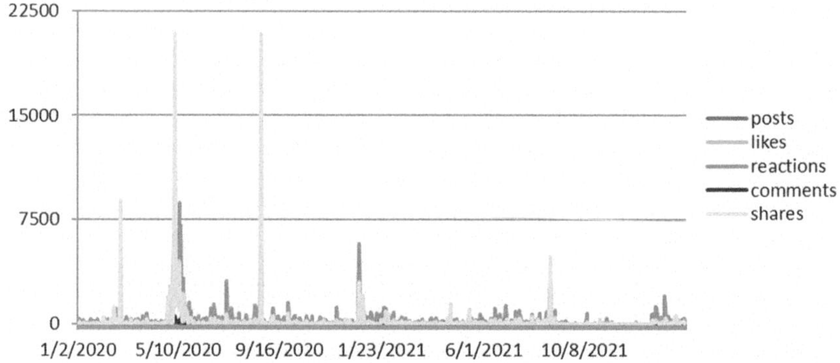

Fig. 2 'Chiloé ta'privao' data *fan page*. (Source: Fan page engagement based on data downloaded with Netvizz)

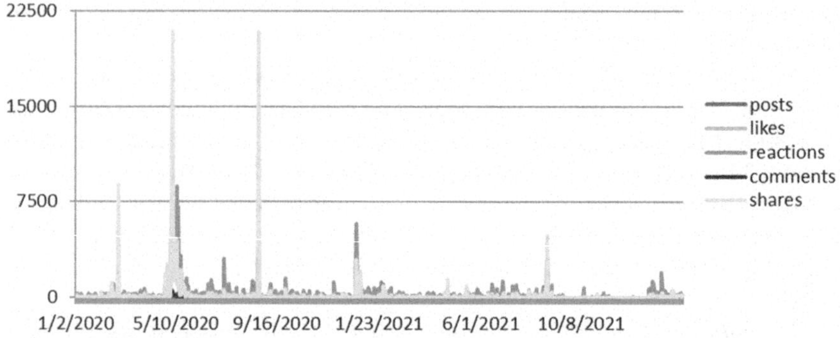

Fig. 3 'Defendamos Chiloé' data *fan page*. (Source: Fan page engagement based on data downloaded with Netvizz)

Discussion

Three Different Strategies Played by the Environmental Activists

To explore local digital activism in the context of the red tide crisis, we have identified three key moments that crystalize the way activists use the Facebook sites to claim their cause. It begins with a single post in early March 2016.

On 7 March 2016, a user named 'Colonode Aguas[7]' [translation: water settler] denounced on a Facebook post the "salmon industry's environmental destruction" of Chiloé. In this text, they claimed that an algal plague was related to the aquaculture industry. A few days later (12 March), the *fan page* administrator published, "We predicted it", when he posted a media report from the local publication, *La Estrella de Chiloé*, warning about the dumping of dead salmon: "Experts caution serious pollution in the sea". This was the genesis for the link between the red tide and salmon industry.

The argument appeared in the three *fan pages* almost on the same days. The dumping of dead salmon was no longer a pollution focus; it was what caused the red tide crisis. While 'Defendamos Chiloé' published a chronology of recent facts, from the dumping to unusual stranding that occurred in Chiloé's beaches since 1 March, the CESCH proposed a technical argument, filming a video where fishermen questioned the damage caused by the dumping, especially the fact that "the red tide doesn't kill bivalves" (CESCH 2016a). The video shows a beach where the spectator can see millions of *machas*[8] over the sand, and then some fishermen explaining why it couldn't be a natural phenomenon. This video had 349 likes and was shared 3113 times.

Then a wave of links, published by legacy media, posted in the analysed *fan pages*. It didn't matter that the link between the dumping and the red tide was not proven scientifically. The argument was installed in the activists' space anyway. For example, on 3 May the 'Defendamos Chiloé' administrator published a screenshot from a TV Channel (TVN) where it was communicated that: "The red tide has spread through the Los Lagos Region[9]" (Defendamos Chiloé 2016a). In the same post, the administrator wrote: "At least! It's not only red tide, fishermen have told that it was caused by the dead salmon dumping on its coasts". This post had 842 likes and was shared 524 times.

Gradually, the links to the legacy media became infrequent. The news coverage followed the social crisis, especially linked to fishermen's labour

[7] This user will be later recognized by the activist as an expert who contested the scientific research and results.

[8] A typical Chilean bivalve.

[9] The Chiloé's region.

conditions: the precariousness of their work, strictly dependent on the sea, leaving hundreds of people with few possibilities to survive. However, the argument that blamed the salmon farming for causing the red tide remained alive in the analysed network and it was spread from different angles.

First of all, the claims that linked the salmon dumping and the red tide persisted, proposing a causal relation between those facts. On 15 May 2016, the CESCH argued on a post: "Could it be affirmed without previous investigation, that the mortality was produced only by the algal blooming of Alexandrium catenella in the Chiloé Sea[10]? Or would it be prudent to thoroughly investigate this and other hypotheses?" (CESCH 2016b). This post had 32 likes and was shared 55 times.

Secondly, the claim mutated to a broader frame where blame for the red tide was not only limited to the dumping itself, but responsibility extended to the whole salmon industry. This argument took many variations and became stronger, drawing upon some previous claims that blamed regional aquaculture for environmental degradations (Buschmann 2005). These variations invited the participation of new stakeholders in the argument's survival, like NGO or experts. For example, in the 'Chiloé ta'privao' *fan page*, a post published by Greenpeace Chile[11] appeared on 6 September 2016. It was a cartoon that represented a salmon executive, dressed in a suit. He was hiding some dead species under the water. A text that accompanied the illustration directly linked the salmon industry and the red tide crisis: "The government authorized the dumping of salmon in the sea. The socio-environmental crisis turned deeper on the island. This is what we discovered after 4 months of independent investigations" (Greenpeace 2016).

Finally, the determination to expel the salmon farming from Chiloé grew on the three *fan pages*. The pollution, the social disintegration and the cultural destruction of the island became regular arguments posted on those spaces. We identified three strategies in the *fan pages* where the activists mobilized their arguments. Both causalities, dumping/red tide and salmon farming/red tide, were observed.

[10] A harmful algal bloom produces the red tide. Among those microorganisms, *Alexandrium catenella* is toxic for human beings.

[11] Founded in 1995, Greenpeace Chile is probably one of the country's most important NGOs. Since its foundation, they are engaged in a long list of socio-environmental conflicts. For more details, see: https://www.greenpeace.org/chile/nuestra-historia/

(a) *Linking the controversy with the newsworthy*

Once the red tide passed, the legacy media lost its interest in the crisis. But the activists posted systematically all the news that were linked with the arguments, in an effort to constantly reignite the controversy in legacy media. Initially, lawsuits were brought against the salmon industry. Many activists complained under the judicial authority against the aquaculture environmental damage. The 'Defendamos Chiloé' *fan page* posted on 26 May 2016 the lawsuits' objectives, where it highlights: "to determinate the existence of environmental crimes" (Defendamos Chiloé 2016b). More than a year later they published the state of those lawsuits, on the same *fan page*. (Defendamos Chiloé 2017a). Later, on 29 December 2017, when the Environmental Court of Valdivia—which has a jurisdiction across southern Chile—rejected a lawsuit against a salmon enterprise because of the dumping, the activists' refusal to accept that decision was published in legacy media[12] and posted on the same *fan page* (Defendamos Chiloé 2018a). The one-year anniversary of the red tide crisis of 2016 motivated several organizations on the island to reignite the controversy. They wrote: "we are coordinating several activities in commemoration of the rebellion from last year, among these activities there will be a press conference, courses, signing of an important agreement and talks in Ancud[13] and Castro of Ecocéanos[14] and Greenpeace" (Defendamos Chiloé 2017b).

The CESCH *fan page* posted a digital media article (Menares Velásquez 2017) when a boat carrying dead salmon sank in front of Chiloé, and it wrote: "It is 200 meters from the coastal area, which is the richest from the point of view of biodiversity and marine production. There are seaweed meadows, for example. In case of any disaster, that would have disastrous consequences for all activity, the ecosystem and the communities near Chonchi" (CESCH 2017). This post had 22 likes and was shared 77 times. Many inhabitants continued to publish in the days after that accident, acting as citizen journalists. For example, one participant posted on his Facebook page on 30 October 2017: "At 11 o'clock today, in the Provincial Government, the meeting was held (...) there were present the

[12] Radio Cooperativa online.
[13] A city in Chiloé.
[14] An NGO.

provincial Governor, maritime, the Sernapesca[15] and Health authorities" (Belmar Russell 2017)—clearly performing the role of a citizen journalist.

(b) *Science with a goal*

The activists' relationship with scientific arguments was far from neutral. A first reaction was against the Red Tide Commission. The government, through its Ministry of Economy, convoked and co-financed an interdisciplinary commission to determine if the dumping of the salmon was related to the red tide phenomenon. The commission published a report (Comisión marea roja 2016) confirming that there were no causalities between those two events. Activists reacted on the analysed *fan pages*. The 'Defendamos Chiloé' *fan page* published a post calling for protest against the report: "On Friday, 18th November, we convened a group of people and leaders in the outskirts of the provincial government to express our displeasure with the conclusions of the scientific report commissioned by the government" (Defendamos Chiloé 2016c).

The broader argument that links the red tide with the entire salmon farming activity was rich soil for argument about science. Experts emerged, presenting contesting studies that affirmed the direct causality of the industrial activity, mobilizing scientific data, like excess nutrients, the sea's load capacity or water eutrophication. An article written by a university professor was posted in the 'Chiloé ta'privao' *fan page*. The document explained, with satellite images and consideration of the water temperature, why the salmon dumping caused the red tide (Trauco Chiloé 2016).

The 'Chiloé ta'privao' *fan page* criticized an unknown University of Valparaíso study (we didn't find any reference of this study) a few days later. This post had 40 likes and was shared 41 times:

> Group of Scientists of the University of Valparaiso – based on satellite photographs – assure that the #MareaRoja[16] is in retreat. Where are the field studies? Does a study based on photos have some seriousness? And what about the other two types of algae that left the sea without oxygen and killed the fauna and released a green foam? (Chiloé ta'privao 2016a)

[15] The National Fisheries and Aquaculture Service of Chile.

[16] Red tide in Spanish.

A YouTube video posted in the 'Defendamos Chiloé' *fan page* said: "Biologist Hector Kol reveals the truth about the red tide crisis" (Maripillán 2016). They started, as well, some Twitter polls (Defendamos Chiloé 2016d), asking its members if they believed the Greenpeace report more or the Red Tide Commission (the Greenpeace alternative won a majority of 90%). The same Greenpeace report motivated a CESCH publication, arguing that "the government betrayed Chiloé, they sacrificed it for the salmon industry". The NGO report indicated that "there was no oceanographic evidence, based on NASA satellite images, to confirm that there was no link between the dumping and the red tide" (CESCH 2016c). This post had 30 likes and was shared 84 times.

The local press[17] was also part of this mobilization of science as an argument. An article was posted on the 'Defendamos Chiloé' *fan page*. Its headlines pointed to the local uncertainty about scientific studies: "Science doesn't have an explanation" and the "Red tide: a new study reveals a strange presence of organic material in the sea". There was also a comment posted by the *fan page* administrator: "You see? We still don't know what really happens! It's a shame for the Red Tide Commission" (Defendamos Chiloé 2016e). This post had 27 likes and was shared six times.

An article published in *Nature* (León-Muñoz et al. 2018) explained with atmospheric, hydrologic and oceanographic data why the water temperature increased in the summer of 2016, and why those natural conditions were linked with the harmful algal bloom observed in that period. The CESCH *fan page* critiqued this position. On a post published on February 2018 that had 22 likes and that was shared 17 times:

> [We think] it is worrying that the last scientific publication on the subject (Leon-Muñoz et al. 2018), states that the February algal bloom (the one that killed the salmon) was only due to the hydro-climatic conditions. This study has a weak methodology, a lack of data exposure, and speculations. To argue that there were only external conditions is worrisome. There is information that confirms the growing relationship between salmon farming (eutrophication), algal flowering and red tide. The salmon impact is not an element to discard. (CESCH 2018)

[17] *La Estrella* (Chiloé's newspaper).

(c) *New metaphors*

We observed many variations in how the activists wrote about its causes. The salmon industry's impact was called 'environmental destruction' (Colono de Aguas 2016a), an 'environmental disaster' (Colono de Aguas 2016b) or 'an environmental earthquake of the salmon' (Defendamos Chiloé 2016f). They claimed that 'this industry is committing a salmon environmental crime' (Defendamos Chiloé 2017c); that 'they are looting the Chilean sea' (Muñoz Ponte 2016), 'destroying it' (Defendamos Chiloé 2018b) and 'converting it to a salmon landfill' (Defendamos Chiloé 2018c); and that 'the Chilean sea is not a rubbish dump' (Chiloé ta'privao 2016b). Regarding the socio-environmental crisis, they mobilized a discourse about the crisis that assembled hundreds of fishermen against the salmon industry and the possible links with the red tide phenomenon. Activists baptized the crisis' one-year anniversary as the 'Tragedy of May' (Defendamos Chiloé 2017d), organizing different events to pressure the state for possible responsibilities.

Conclusions

The intensity associated with outbreaks of red tide shows the different strategies employed by the activists. In order to make this observation, we could establish certain guidelines around the publication and reproduction of content. The possible links between the red tide and the salmon dumping is part of a major movement: the local activists against the salmon aquaculture industry. As we proposed in our methodology, the social actors that were at the origin of the controversy continued talking about the salmon farming and the red tide after the natural phenomenon passed. In that sense, its activity in the networks allows the controversy about the possible link between red tide and the salmon industry to remain open, at least on these Facebook sites. On the other hand, it also shows that digital media continue to distribute news on the subject long after legacy media lost interest. As long as there is still talk of salmon farming and its negative impact for the region, the issue remains in force and contributes to the formation of a highly social discontent ground against situations of environmental risk. One of the roles of these Facebook sites is to maintain interest and, in particular, to sustain the perceived environmental threat of salmon farming. Each time a controversy arises again, local networks can be a source for the legacy media. This symbiotic relationship between local digital networks and legacy media is substantive in understanding environmental conflict today.

In a longer perspective (2016–2018), our analysis of Facebook's posts brings to light some of the roles Cammaert's (2015) identified: the importance of social media to disseminate frames independently of the main currents and surveilling the surveilles. The local activism against salmon farming existed before the red tide of 2016, and it continued after the natural phenomenon. The controversial frame was adapted to their cause, maintaining legacy media interest, and it is highly reactive and efficient for eventual protest organization.

The permanent activity responds to a need of content that is provided by the media (traditional or digital), linked directly with the news. Activists publish content from digital media when there's no news from the legacy media linked to its causes. We observed occasionally that digital media appeals to the activist social media when there are no official sources covering socio-environmental conflicts. Monitoring the work of activists influences legacy media activity when there are no other 'reliable' sources and also serves to generate identity and cohesion for the group. The networking of these activists represents, then, a way to implement a permanent strategy of communication action in the absence of greater interest from legacy media in these conflicts.

This chapter explores how digital activism in the local context contributes to the survival of a controversy. In fact, from this time, we observe important and systematic activities of digital activists (Millaleo and Velasco 2013) in the social networks. They engaged the controversy, shifting their focus of interest to combat salmon activity in the region. By associating networks of activists on social media, we demonstrate that these platforms can be a way to continue with the dynamics of the controversy, always trying to influence the agenda of legacy media. The activists' movements that we described in our empirical work demonstrate how the locals met a national—and a global—context in the red tide crisis. Coverage of the socio-environmental controversy by the legacy media could be interpreted as an opportunity for local people to show to other environmental activists and populations about the salmon industry impacts whereby the contest between local digital networks and often-distant legacy media jostle to limit and expand public debate and understanding of an environmental issue.

References

Becerra, M., & Mastrini, G. (2017). *La concentración infocomunicacional en América Latina. Nuevos medios y tecnologías, menos actores*. Bernal: Universidad Nacional de Quilmes.

Belmar Russell, P. A. (2017). Crisis de Pilpilehue y todo Chiloé. Defendamos Chiloé's Facebook fan page. https://www.facebook.com/472249059604015/posts/860460217449562. Accessed 13 Oct 2018.

Bennett, W. L., & Segerberg, A. (2012). The logic of connective action. *Information, Communication & Society, 15*(5), 739–768.

Bresnahan, R. (2003). The media and the neoliberal transition in Chile: Democratic promise unfulfilled. *Latin American Perspectives, 30*(5), 39–68.

Broitman, C., & Kreimer, P. (2018). Knowledge production, mobilization and standardization in Chile's HidroAysén case. *Minerva, 56*(2), 209–229.

Buschmann, A. (2005). *Marea roja y salmonicultura en el sur de Chile*, Oceana. Available in http://www.bio-nica.info/biblioteca/Buschmann2005.pdf

Cammaerts, B. (2015). Social media and activism. In R. Mansell & P. Hwa (Eds.), *The international encyclopedia of digital communication and society* (pp. 1027–1034). Oxford, UK: Wiley-Blackwell.

Castells, M. (2007). Communication, power and counter-power in the network society. *International Journal of Communication, 1*(1), 29.

Centro de estudios sociales de Chiloé. (2016a). ¿Solo es marea roja ? CESCH's Facebook fan page. https://www.facebook.com/1507235036167576/posts/2009012785989796. Accessed 13 Oct 2018.

Centro de estudios sociales de Chiloé. (2016b). Otros antecedentes a considerar sobre la crisis ambiental en el mar de Chiloé. CESCH's Facebook fan page. https://www.facebook.com/1507235036167576/posts/2018418441715897. Accessed 13 Oct 2018.

Centro de estudios sociales de Chiloé. (2016c). Greenpeace: el gobierno traicionó a Chiloé, lo sacrificó en pro de la industria salmonera. CESCH's Facebook fan page. https://www.facebook.com/1507235036167576/posts/2091143074443433. Accessed 13 Oct 2018.

Centro de estudios sociales de Chiloé. (2017). Está a doscientos metros del area costera. CESCH Facebook fan page. https://www.facebook.com/1507235036167576/posts/2394219534135784. Accessed 13 Oct 2018.

Centro de estudios sociales de Chiloé. (2018). En estos días volvemos a hablar de bloom algal y de marea roja. CESCH's Facebook fan page. https://www.facebook.com/1507235036167576/posts/2463756693848734. Accessed 13 Oct 2018.

Chiloé ta'privao. (2016a). Grupo de científicos de la Universidad de Valparaíso. https://www.facebook.com/1000694046684798/posts/1010149659072570. Accessed 13 Oct 2018.

Chiloé ta'privao. (2016b). Gracias estudiantes! Nuestro mar no es basurero de los empresarios. Chiloé's ta'privao's Facebook fan page. https://www.facebook.com/1000694046684798/posts/1001482119939324. Accessed 13 Oct 2018.

Colono de Aguas. (2016a). Chiloé enfrentado a plaga de algas y a su destrucción ambiental por la industria del salmon. Defendamos Chiloé's Facebook fan page. https://www.facebook.com/472249059604015/posts/1010740555638235. Accessed 13 Oct 2018.

Colono de Aguas. (2016b). la corrupción que dejó al desnudo el desastre ambiental de Chiloé. CESCH's Facebook fan page. https://www.facebook.com/1507235036167576/posts/2023584321199309. Accessed 13 Oct 2018.

Daniels, J., & Mather, C. (2017). Conservation and care: Material politics and Atlantic salmon on Newfoundland's Gander River. *Maritime Studies, 16*. https://doi.org/10.1186/s40152-017-0072-6.

Davidson, K., Gowen, R. J., Harrison, P., Fleming, L. E., Hoagland, P., & Moschonas, G. (2014). Anthropogenic nutrients and harmful algae in coastal waters. *Journal of Environmental Management, 146*, 206–216. https://doi.org/10.1016/j.jenvman.2014.07.002.

Defendamos Chiloé. (2016a). Por fin!!! TVN ahora: "no solo es marea roja. Los Pescadores señalan que hubo vertimiento de salmon muerto en sus costas". Defendamos Chiloé's Facebook fan page. https://www.facebook.com/472249059604015/posts/577148632447390. Accessed 13 Oct 2018.

Defendamos Chiloé. (2016b). Atención Chiloé! Siguen abiertos plazos para sumarse a querellas por vertimiento de salmon descompuesto. Defendamos Chiloé's Facebook fan page. https://www.facebook.com/472249059604015/posts/586487911513462. Accessed 13 Oct 2018.

Defendamos Chiloé. (2016c). Mañana reunion en Castro. Defendamos Chiloé's Facebook fan page. https://www.facebook.com/472249059604015/posts/673606352801617. Accessed 13 Oct 2018.

Defendamos Chiloé. (2016d). Vota en nuestro Twitter! 90% le cree a Greenpeace. Nosotros también! Defendamos Chiloé's Facebook fan page. https://www.facebook.com/472249059604015/posts/633140470181539. Accessed 13 Oct 2018.

Defendamos Chiloé. (2016e). ¿Ven? Seguimos sin saber qué pasó. VERGONZOSO para la U de los Lagos y su comité de científicos. Defendamos Chiloé's Facebook fan page. https://www.facebook.com/472249059604015/posts/675697325925853. Accessed 13 Oct 2018.

Defendamos Chiloé. (2016f). Muchísimas gracias a la gente que no está colaborando en tema terremoto ambiental salmonero. Defendamos Chiloé Facebook fan page. https://www.facebook.com/472249059604015/posts/555139821314938. Accessed 13 Oct 2018.

Defendamos Chiloé. (2017a). Aquí podrás ver como revisar estado de las 2000 querellas impuestas el año pasado por vertimiento de salmones. Defendamos

Chiloé's Facebook fan page. https://www.facebook.com/472249059604015/posts/805940486234869. Accessed 13 Oct 2018.

Defendamos Chiloé. (2017b). Próxima conmemoración de rebellion socio-ambiental de Chiloé. Defendamos Chiloé's Facebook fan page. https://www.facebook.com/472249059604015/posts/754025551426363. Accessed 13 Oct 2018.

Defendamos Chiloé. (2017c). El viernes mandaremos un aviso a Chile y el mundo: somos patrimonio de la humanidad, no aceptaremos más abusos ni crimen ambiental salmonero. Defendamos Chiloé Facebook fan page. https://www.facebook.com/472249059604015/posts/729844463844472. Accessed 13 Oct 2018.

Defendamos Chiloé. (2017d). Organizaciones sociales, ambientales y pueblos originarios comenzamos a conmemorar la tragedia de mayo. Defendamos Chiloé's facebook fan page. https://www.facebook.com/472249059604015/posts/757791847716400. Accessed 13 Oct 2018.

Defendamos Chiloé. (2018a). Partimos el 2018. Defendamos Chiloé's Facebook fan page. https://www.facebook.com/472249059604015/posts/897935317035385. Accessed 13 October 2018.

Defendamos Chiloé. (2018b). Las salmoneras ya tienen destruido el mar en Chiloé, ya están en Aysén y van por Magallanes. Defendamos Chiloé's Facebook fan page. https://www.facebook.com/472249059604015/posts/922340161261567. Accessed 13 Oct 2018.

Defendamos Chiloé. (2018c). Aysén y Chiloé no pueden seguir siendo el vertedero salmonero. Defendamos Chiloé's Facebook fan page. https://www.facebook.com/472249059604015/posts/933832693445647. Accessed 13 Oct 2018.

Della Porta, D., & Diani, M. (2011). *Los movimientos sociales.* Madrid: Centro de Investigaciones Sociològicas UCM.

Geoffroy, E., & Sunkel, G. (2001). *Concentración económica de los medios de comunicación.* Santiago: LOM Ediciones.

Godoy, S., & Gronemeyer, M. E. (2012). *Mapping Digital Media: Chile. A report by the Open Society Foundations.* https://www.opensocietyfoundations.org/reports/mapping-digital-media-chile. Accessed 13 Oct 2018.

González-Bustamante, B., & Soto Saldías, D. (2016). Pluralismo en los medios digitales escritos en Chile, una exploración de casos clave. *Serie Política, Gestión y Políticas Públicas, 2*, 1–26.

Greenpeace. (2016). Chiloé: que no te escondan la verdad. Chiloé ta privao's Facebook fan page. https://www.facebook.com/1000694046684798/posts/1082091098545092. Accessed 13 Oct 2018.

Gronemeyer, M. E. (2013). La digitalización y sus efectos sobre las prácticas y productos periodísticos en Chile. *Palabra Clave, 16*(1), 101–128.

Gudynas, E. (2014). Los extractivismos en sus contextos globales, regionales y Locales. In *Memoria Seminario Internacional Extractivismo en América Latina ... Agua que no has de beber* (pp. 11–23). OLCA- OCMAL: Santiago.

Harrison, H. L., & Loring, P. A. (2014). Larger than life: The emergent nature of conflict in Alaska's upper cook inlet salmon fisheries. *SAGE Open*. https://doi.org/10.1177/2158244014555112

León-Muñoz, J., Urbina, M. A., Garreaud, R., et al. (2018). Hydroclimatic conditions trigger record harmful algal bloom in western Patagonia (summer 2016). *Scientific Reports, 8*, 1330. https://doi.org/10.1038/s41598-018-19461-4.

Lester, L., & Hutchins, B. (2012). The power of the unseen: Environmental conflict, the media and invisibility. *Media, Culture and Society, 34*(7), 847–863.

Lim, M. (2013). Framing Bouazizi: 'White lies', hybrid network, and collective/connective action in the 2010–11 Tunisian uprising. *Journalism, 14*(7), 921–941.

Mansour, E. (2012). The role of social networking sites (SNSs) in the January 25th Revolution in Egypt. *Library Review, 61*(2), 128–159.

Maripillán, D. (2016). Explicación de biólogo marino por contaminación de nuestro mar. Chiloé ta privao's Facebook fan page. https://www.facebook.com/472249059604015/posts/10154109992014356. Accessed 13 Oct 2018.

Menares Velásquez, F. (2017). Una bomba de tiempo. Advierten riesgo ambiental del barco hundido con salmons podridos en Chonchi. *El Ciudadano*. https://www.elciudadano.cl/medio-ambiente/una-bomba-de-tiempo-advierten-riesgo-ambiental-del-barco-con-salmones-podridos-hundido-en-chonchi/10/30. Accessed 13 Oct 2018.

Millaleo, S., & Velasco, P. (2013). *Activismo digital en chile. Repertorios de contención e iniciativas ciudadanas*. Santiago: Fundación democracia y desarrollo.

Muñoz Ponte, N. (2016). Orgullosa del pueblo chilote que defiende su patrimonio atural con tanto ahínco!! Defendamos Chiloé's facebook fan page. https://www.facebook.com/472249059604015/posts/10209415133398020. Accessed 13 Oct 2018.

Noakes, D. J., Fang, L., Hipel, K. W., & Kilgour, D. M. (2003). An examination of the salmon aquaculture conflict in British Columbia using the graph model for conflict resolution. *Fisheries Management and Ecology, 10*(3), 123–137.

Osmundsena, T. C., & Olsen, M. S. (2017). The imperishable controversy over aquaculture. *Marine Policy, 76*, 136–142.

Ramírez Cáceres, J. D. (2010). Radios Comunitarias en Chile: las paradojas de su propiedad. *Revista Austral de Ciencias Sociales*, (19), 63–74.

Reyes Herrera, S., & Rodríguez Torrent, J. C. (2015). Proyecto Hidroaysén: capitalismo extractivista, regulación estatal y acción colectiva en la Patagonia. *Polis, 4*(40), 439–467.

Sáez Baeza, C. (2013). Políticas públicas de comunicación y participación ciudadana: el caso de la televisión digital en Chile. *Signo y Pensamiento, 32*(63), 34–51.

Sapiezynska, E. (2014). Los No Aparecidos: la Protesta Social Invisible en los Grandes Medios en Chile y las Políticas Mediáticas del Disenso. *Comunicación y medios, 29*, 156–170.
Singleton, S. (2009). Native people and planning for marine protected areas: How "stakeholder" processes fail to address conflicts in complex, real-world environments. *Coastal Management, 37*(5), 421–440.
Trauco Chiloé. (2016). Derribando hipótesis de las salmoneras y el Estado. Chiloé ta privao's Facebook fan page. https://www.facebook.com/1000694046684798/posts/984182061636270. Accessed 13 Oct 2018.
Urkidi, L. (2010). A glocal environmental movement against gold mining: Pascua–Lama in Chile. *Ecological Economics, 70*, 219–227.
Valenzuela, S., Arriagada, A., & Scherman, A. (2012). The social media basis of youth protest behavior: The case of Chile. *Journal of Communication*. https://doi.org/10.1111/j.1460-2466.2012.01635.x.
Valenzuela, S., Arriagada, A., & Scherman, A. (2014). Facebook, twitter, and youth engagement: A quasi-experimental study of social media use and protest behavior using propensity score matching. *International Journal of Communication, 8*, 2046–2070.
Wallerstein, I. (1998). *El moderno sistema mundial I*. México: Siglo XXI.
Wolfsfeld, G., Segev, E., & Sheafer, T. (2013). Social media and the Arab Spring: Politics comes first. *The International Journal of Press/Politics, 18*(2), 115–137.
Youmans, W. L., & York, J. C. (2012). Social media and the activist toolkit: User agreements, corporate interests, and the information infrastructure of modern social movements. *Journal of Communication, 62*(2), 315–329.

ICTs, Environmental Activism and Community Mobilization in Senegal, Kenya and South Africa

Tracy M. Tinga, Patrick D. Murphy, and Emmanuel Septime Sessou

The Global South often serves as the barometer for Earth's climate change crisis, evidenced through a host of environmental riskscapes, made salient in relation to food security, public health, natural resource stewardship, species extinction, and poverty. This is not because the South has necessarily lagged behind in the race for modernization, but rather, as South African media scholar Herman Wasserman (2018) argues, "it is in many ways ahead of the curve of history—that, in the contemporary, global world, it is 'regions in the South that tend to first feel the concrete effects

T. M. Tinga (✉)
Media and Communication Studies, University of Maryland, Baltimore County, Baltimore, MD, USA

P. D. Murphy • E. S. Sessou
Klein College of Media and Communication, Temple University, Philadelphia, PA, USA
e-mail: murphy.p@temple.edu; eseptime.sessou@temple.edu

J. Díaz-Pont et al. (eds.), *The Local and the Digital in Environmental Communication*, Global Transformations in Media and Communication Research - A Palgrave and IAMCR Series,
https://doi.org/10.1007/978-3-030-37330-6_7

of world-historical processes as they play themselves out, thus to prefigure the future of the former metropole.'" He continues, "if you want to find out what the effects of climate change will be in the North, speak to farmers in African villages who are already struggling with its impact on their livelihoods" (p. 56).

In the case of sub-Saharan Africa, the focus of this chapter, global processes of political and economic transition, whose promises ostensibly laid with "development" and "progress," have too often exposed the less celebrated indicators of globalization, namely environmentally rapacious practices ranging from, for instance, the devastating impact of the petroleum industry extractivism, the vast global dumping e-waste, and the illegal trade in rhino horns, elephant tusks, and other "products" stripped from the most iconic symbols of African environmental identity. But despite their undeniable presence, until recently drawing public attention to environmental injustices has often been difficult, especially given the region's history of authoritarian regimes, lack of communication infrastructure, tightly controlled messaging, and low literacy rates, assuring that it long remained one of the world's most ecologically vulnerable regions. However, this media landscape has changed quite dramatically over the course of the last several decades, and with it the possibilities for engaged ecological citizenship.

As communication scholars have shown, many sub-Saharan African countries have actually moved with alacrity into the digital era, with the introduction of new information and communication technologies (ICTs) being hailed as vital for the promotion of economic development, for improving government and NGO effectiveness, and for the promotion of peace (Garliardone et al. 2015). While there is a solid corpus of scholarship on how these changes are altering the political landscape, little has focused on how ICTs have been used to motivate the public to respond the mounting environmental concerns. In an attempt to begin to address this gap in the research, this chapter examines how the adoption of ICTs in sub-Saharan Africa is reshaping how citizens are able to respond to the challenges of climate change and related environmental issues by opening up opportunities for more regionalized and localized community building, knowledge sharing, and action.

The chapter begins by placing the emergence of ICTs within the broader sub-Saharan media terrain through an overview of how, from the mid-1990s to today, media systems have been transformed by external and internal forces associated with neoliberal globalization which has in turn profoundly altered how African's receive, share, and create media content.

The chapter continues by explicating how, more recently, ICTs have reconstituted African mediascapes, particularly in terms of how collaborative platforms and social media have given citizens greater voice and the capacity to tell their own stories and, by implication, the possibility to define and address their environmental realities.

To explore these changes at a more granular level, the chapter then offers three short case studies as illustrations of how the adoption of ICTs has engendered citizen-driven environmental communication. The first case presents VUMA.EARTH, an online campaign website launched by Greenpeace Africa, which advocates for the use of peaceful and creative confrontation to expose environmental injustices. Designed to empower citizens with the creation of call-to-action environmental campaigns, the website provides the tools, "how to" guidelines, and online platform to help African citizens to target and respond to everything from hyper local problems (e.g., e-waste in particular city dump) to cross-border campaigns tackling broader ecological challenges (e.g., banning single use plastic). The second case examines how farming communities in rural Kenya are using mobile phone apps to establish information networks to strategically adapt to the challenges of climate change and its effects on agriculture as a livelihood. The emphasis in this case study is placed on knowledge acquisition and problem-solving, as community members share strategies for everything from livestock management among pastoralist communities and information for precision farming, to market information. The third case presents the background behind Water 4 Cape Town, a social media water stewardship campaign initiated by a single South African resident and that grew into highly networked communication effort designed to engage marginalized communities in resource management and green lifestyles through grassroots strategies, with water as the entry point towards sustainable environmental practice. As these cases reveal, ICTs have permitted citizens to mobilize in new ways through the elaboration of local and regional virtual communities and knowledge sharing to identify and tackle—not just become aware of—ecological problems.

The chapter concludes by digesting the lessons of these examples of environmental communication in relation to sub-Saharan Africa's fertile media landscape. Specifically, it considers how local actors have used the affordances of ICTs to increase environmental stewardship through awareness campaigns and knowledge sharing networks, but also in terms of political engagement—a transformative process consistent with recent calls to "politicize" the environment through communication that fosters public action (Carvalho et al. 2017).

The Sub-Saharan Media Landscape and the Emergence of Digital Communities

It is important to understand that the emergence of ICTs in sub-Saharan Africa has taken place within national mediascapes that had been profoundly reshaped by neoliberal globalization. Within this context, the notion of modernization—an idea linked back to an earlier, post-World War II development agenda—was recast in terms of liberalization, privatization and democratization (Harrison 2013; Murphy 2017). To make sense of these changes, much scholarship on sub-Saharan media systems has focused on how a combination of external and internal forces reconfigured the way that African's receive, share, and even participate in the creation of media content (Avle 2011; Blankson 2005, 2007; Eko 2003; Emeagwali 2011; Musa 2011; Nyamnjoh 2013; Oduro-Frimpong 2014; Onwumechili 2007; Sparks 2009; Wasserman 2011, 2018). As this corpus of scholarship shows, externally triggered waves of "good governance" programmes pressured many African nations to privatize their national media systems to attract foreign investments. The push to privatize was in part driven by the logic that liberalization would catalyse democratization once the control of the media was out of the hands of authoritarian regimes. Internally, the push for change came from different social actors such as student groups and educators, trade unions, religious organizations, and, even more recently, media makers themselves. These actors sought greater participation in the public sphere (Blankson 2005, 2007; Grätz 2011, 2013; Nyamnjoh 2013).

Within the confluence of these forces, sub-Saharan mediascapes began to position cultural modernity and political participation as byproducts of marketization, providing a foundation for the notion that cultural expression and social status, among other things, were naturally aligned with the adoption of consumer lifestyles (Blankson 2005; Harrison 2013; Oduro-Frimpong 2014). Significantly these forces of change also activated a more mobilized citizenry seeking political agency and voice in concert with a hunger for entrepreneurial opportunities. The growth of localized and regionalized media entrepreneurship in the form of both legacy media (radio and TV/video) and emergent media followed, presenting in the process new spaces for a greater diversity of voices, and increased possibilities for civic engagement (Avle 2011; Grätz 2011, 2013; Musa 2011; Nyamnjoh 2013; Thalén 2011).

Sub-Saharan Africa's liberalized commercial turn also made the region attractive to a new set of players (Murphy 2017). Most notably, starting in the 2000s China began to work hard to insert itself in multiple sub-Saharan nations through infrastructure-based development projects, but also more strategically, through "soft power" interventions. This involved investing heavily in sub-Saharan Africa's emerging communication networks, even establishing state-owned news media offices such a CCTV, China Radio International, and the Xinhua news agency in Kenya and South Africa (Wasserman 2018, p. 137). China framed its application of soft power as driven by its broader anti-colonial and anti-imperialist agenda, yet this investment has brought into question whether these efforts are not just another means to foster cultural reproduction and dependency (Banda 2009). However, as Tinga (2019) asserts, the "partnership" between China and various African countries is also seen as challenging the hegemonic relationship between Western financial institutions (e.g., the World Bank and IMF) and African countries. This view is tied to the fact that China does not stipulate the economic policies it expects the African countries to adhere to, but rather understands this relationship more as a trade partnership, which has given rise to the notion of "South-South Cooperation (SSC)"—an idea defined as being mutually beneficial, based on mutual friendship, goodwill, equality, and sharing of knowledge (King 2013).

China's objectives notwithstanding, in most instances the "new players" have been Africans themselves (Murphy 2017). Pan-African media markets have surfaced in relation to West African video and film industries, often with direct ties to the intertwined Nigerian-based "Nollywood" and the Ghanaian "Gollywood," which produce a quantity of films surpassed only by Bollywood (India) and Hollywood (the US) (Krings and Okome 2013; Oduro-Frimpong 2014). More locally, liberalization and privatization have also triggered the resurgence of regionalized radio, the medium of choice for most Africans; a rebirth sparked in no small way to its capacity to draw sponsors from the now expanding advertising industry (Grätz 2013). Moreover, in an effort to create attractive audience packages for large corporate multinational sponsors, large production houses from various countries have hired African "entertainment brokers" to established pan-African markets for their television productions (Thalén 2011). These developments have shaped the contours of a vibrant media ecology and a rising sense of "Afro-cosmopolitanism" elaborated and maintained by a whole class of new African cultural actors (e.g., filmmakers, journalists, advertisers, disc jockeys) (Avle 2011; Grätz 2011, 2013; Musa 2011; Tinga 2019).

The most recent wave of change has taken place via the adoption of new information and communication technologies (ICTs), which many sub-Saharan nations have been quick to embrace (Mudhai et al. 2009). Early on the affordances of ICTs were heralded as a means to leapfrog social change impediments and thus became a key topic within development circles because ICTs were considered central to the implementation of social change initiatives (Manyika et al. 2013; Unwin 2009). Echoing the past history of development discourses, the notion that ICTs could help nations modernize was championed in the belief that they would bridge the digital divide, enabling developing countries that lagged behind technologically to "catch up" (Kleine and Unwin 2009). This digital push has led national governments to adopt policies that put ICTs at the heart of their development agendas and companies that manufacture technologies capitalizing on this, resulting in what was called the "digital revolution" (Kleine and Unwin 2009). Subsequently, many governmental agencies and NGOs moved to ICT-driven communication models, as it is through the affordances of ICTs that governments have positioned themselves to serve their citizens through e-governance (Mudhai et al. 2009), and international actors (World Bank, IMF, NGOs) to engage and share local knowledge to inform development initiatives (Garliardone et al. 2015).

Despite the more patronizing tones of this ICTs adoption agenda and charges that they have been presented as a technologically deterministic panacea to underdevelopment, the significance of ICTs in sub-Saharan African countries cannot be downplayed. Indeed, the proliferation of ICTs and the affordances they present to African publics has enriched already-fertile ground for knowledge sharing, civic engagement, and the emergence of diverse voices, by providing citizens with the spaces and tools to produce and distribute their own content. While this opening of and access to "the virtual" has largely been aligned with opportunities for social change and political agency (framed in terms of economic growth through entrepreneurship, education, e-health, e-governance, and rural development (Unwin 2009)), it can also be understood in terms of African's speaking on their own behalf, as witnessed through hashtags such as #theafricatheynevershowyou, #everydayafrica or blogs, vlogs, and social media content that aim at showing the normalcy of life in various parts of the continent and that narrate the different ways that Africans participate in their own advancement (Tinga 2019). Interestingly, even as these capacities for resilience and adaptability align with neoliberal discourses of agency expressed through the market (Chandler and Reid 2016), such creative signs of ICTs application also echoes one of the goals of the New World

Information and Communication Order (NWICO) debates, where developing nations *tell their own stories* (Gumucio and Tufte 2006). In short, it has been through ICTs that Africans have been able to connect the local with the global, and vice versa (Garliardone et al. 2015; Ogan et al. 2009).

ICTs, Participation and the Possibilities of Environmental Activism

Given the vibrancy of the sub-Saharan African mediscapes, particularly in relation to the collaborative potentialities of ICTs, it is perhaps no surprise that many citizens and communities have developed virtual networks to adapt to local and regional environmental challenges. In broad and practical terms, these have been established and utilized along the lines of what Pezzullo and Cox (2017) define as the central role of environmental communication, which is to "alert" (identify and name key issues), "amplify" (broaden awareness and understanding of the issues), and "engage" (move to action) citizens to ecological issues. The potential for ICTs to drive climate change social change efforts has been recognized by development and environmental communication practitioners (Lester and Hutchins 2013; Moser and Dilling 2007; Murphy and Tinga 2019), but, at a more explicitly political level, ICTs' use in parts of the sub-Saharan also suggests the potential for citizens to "politicize" environmental communication (Carvalho et al. 2017) by empowering communities, amplifying counter-hegemonic voices, and facilitating the creation of horizontally connected environmental movements capable of contesting the status quo. As the three examples from Senegal, Kenya, and South Africa detailed below demonstrate, ICTs' environmental initiatives are often started by highly motivated individuals (activists) or at-risk communities, but unfold in relation to partnerships with state and/or international actors—a mix of interests and activities that directly reflect the region's liberalized and transforming social terrain.

Campaigning for Environmental Action: VUMA.EARTH in Senegal

VUMA.EARTH is an initiative that crosses multiple borders on the African continent. It is not a single movement originating from a particular environmental or natural resource management crisis. It is rather an extension of Greenpeace Africa, an environment stewardship organization that identifies itself as a growing movement of people acting for the sake of protecting the environment (Greenpeace Africa 2018). The African chapter follows Greenpeace's mission of organizing campaigns designed to activate peaceful and creative confrontation to expose environmental injustices around the world and develop solutions for a green and peaceful future (https://www.greenpeace.org/africa/en/). As such, VUMA.EARTH is driven by a desire to move beyond merely creating awareness or engendering advocacy, aiming instead to help citizens take on activist roles as environmental defenders to promote "positive social change."

VUMA.EARTH was launched by Greenpeace Africa on June 5, 2018, while celebrating World Environment Day in Johannesburg. The website is designed to enable ordinary citizens to start and run, by themselves, online and offline campaigns on any environmental issue affecting their various communities. The functionality of VUMA.EARTH makes it possible for its users to set up petitions, engage and communicate with their supporters, and finally organize offline events and activities. It also offers campaigning tips and networking opportunities. The name of the platform was purposefully chosen, as VUMA is a phrase with interesting meanings in both Swahili and Nguni languages (Greenpeace Africa 2018). In Swahili, which is widely spoken in East Africa, the term means "to spread like wildfire." However, in Nguni languages (a set of 11 Bantu languages of southern Africa), it means "to be in an agreement" (Vuma.Earth 2019). Although the term means different things to various groups of people, VUMA relates to the idea of wildfire and togetherness: two concepts that are particularly relevant to the need of preserving the environment. Encapsulating both "awareness" and "consensus" (Vuma.Earth 2019), it can be interpreted as a call to collective action against the flames of environmental injustice that are devouring and drying up natural resources.

In terms of the geographic scope, campaigns started using the VUMA.EARTH website can be either specifically localized, or national, or even "global" without a given location. The form for creating an online campaign is organized around nine categories, including civil society, climate change, coal, ecological farming, energy, forests, Genetically Modified Organisms (GMO), oceans, and water (Vuma.Earth 2019). Under these groupings, environmental activists have created campaigns in English or French, depending on the country targeted. In this respect, the platform reports hundreds of online campaigns relative to ecological issues in multiple parts of Africa. Countries where campaigns have been started by citizens include Cameroon, Democratic Republic of Congo (DRC), Kenya, Nigeria, Malawi,

Senegal, South Africa, Tanzania, and even as far as Mauritius. This broad set of countries attests to both the increased access to and the affordances of information and communication technologies in sub-Saharan Africa, underscoring how ICTs are bridging the gaps that are imposed by national and geographic borders to mobilize communities around an ideal.

The platform offers tips on how to run and draw users to campaigns, and also presents campaigners with the capacity to collect signatures offline on paper as a way of doing what Costanza-Chock (2012) calls "transmedia mobilization." There's also the possibility of sharing the campaign on social media platforms like Facebook and Twitter where activists and participants can comment and engage in conversations relative to reasons for signing the petition. Some of the campaigns launched on VUMA. EARTH have run for months and obtained as much as 15,616 individual signatures. Those that have drawn the most attention include noise pollution and trophy hunting, as well as multiple regionalized campaigns aimed at banning single use plastic by specific corporations (e.g., Dansi bottled water, Nado's condiments, Wimpy restaurants).

As an example, one of these campaigns has been started to confront e-waste in Senegal, a problem directly related to local-national trade practices. A coastal country of West Africa, Senegal benefits from its maritime economy, as the country has four ports with Dakar as the main one. These assure 90% of Senegalese foreign trade with thousands of containers transiting through its site every year. One of the sectors that has flourished in the past decade is the business of secondhand electronics trade. A used TV, smartphone, or computer turns out to be less expensive when bought out of the port than new ones sold in shopping centres. As a result, the people in Senegal prefer purchasing these electronics that are affordable, but yet typically have a shorter shelf-life. When these devices no longer work, they are sent to the dumps where they are broken or burned in search of precious metals. The consequences are multiple. Not only are toxic gases released in the atmosphere, but the people burning these items or living around the dumps are exposed to respiratory diseases.

The phenomenon has been going for years, especially with Senegalese of the diaspora starting and nurturing transnational businesses in the sector. Mamadou Kaly Ba, an activist who has been campaigning against e-waste in Senegal, thinks that this habit became "a culture in the country, as people prefer buying secondhand electronics because they are cheaper" (Ba, personal communication, May 1, 2019). For him, this is a problem that is ignored by everyone, especially the government and even ecologists. The later ones privilege dealing with deforestation or overfishing issues, while the former does not feel any need to acting against it since taxes are taken out of the transactions. Under these conditions, Mamadou Ba found himself alone and in need of a community to help him address the issue. So, he started a campaign on VUMA.EARTH. The campaign is aimed at 400 signatures as the first step before moving forward. According to Ba, 400 is the number of signatures needed to reach for Greenpeace to take your campaign seriously. Once this number is collected, the organization feels that there is enough leverage to bring the subject before local authorities, who in turn who will feel more compelled to act once they see the campaign is endorsed by Greenpeace Africa (Ba, personal communication, May 1, 2019).

There have been many challenges and lessons learned, however. Ba started the campaign and obtained signatures via WhatsApp groups, Facebook, Twitter, and through emails to his friends and activist colleagues, who in turn shared his petition. He has also collected signatures offline on paper, but this has been difficult. In fact, most of the people he talks to are in the practice of purchasing secondhand electronics, and so understand the hypocrisy of acting against something they themselves do. Ba is also keenly aware that, unlike used car importation in Senegal where car companies have been directly involved in shaping policy, the fight against e-waste may not gain traction because no brand or corporate entity is taking on the fight (Ba, personal communication, May 1, 2019). When social matters need to be interest-bound for the government and corporations make changes, even human life and natural resources are relegated to market forces. As a result, the people are left alone to pick up the fight. But currently under these conditions, change is less likely to happen because it is dependent upon engaged subjects, which reveals a conundrum shaped by the underlying tension between those who see themselves as both consumers and stewards of the environment, which for now is hobbling real environmental action.

Apps for Adaptation: Mobile Phones and Sustainable Agriculture in Kenya

Kenya has been dubbed Africa's "Silicon Savannah" as a reflection of the penetration and growth of digital technologies, more specifically mobile phones. According to the Kenya mobile report 2019 by Jumia Kenya,

Kenya has a 91% mobile subscription penetration and 84% Internet users. These numbers are indicative of the growth of smartphone ownership, which has been catalysed by the entrance of smartphones from Chinese technology firms such as Techno. While some scholars have positioned the growth of mobile technologies in Kenya and other African countries as a panacea to underdevelopment (Aker 2011; Aker and Mbiti 2010; Unwin 2009), others point to the revolutionizing aspect of mobile phones. These include how mobile phones have provided a means for Africans to relate to themselves, to each other, understand themselves better even as they connect with the rest of the world (Mbembé 2001), or how mobile phones have opened up spaces for "e-democracy" by providing citizens with information, campaigning, or monitoring of elections (Wasserman 2011, 2018). There are also more transactional and security-related developments, as mobile money transfer platform M-Pesa that has transformed how people conduct business (Edwards et al. 2016) and the invention of Ushahidi, a crowdsourcing app launched during Kenya's post-election violence in 2007 to map volatile areas (Okolloh 2009). Whatever the case, it is clear that the growth of mobile technologies in Kenya and the affordances it has created for various facets of daily life cannot be ignored.

Apart from these, mobile technologies have also been leveraged to promote climate change adaptation strategies among various communities. Three mobile apps in particular—AfriScout, M-Farm, and Ujuzi Kilimo—have been created to promote sustainable farming practices to mitigate the effects of climate change. The use of these apps illustrates characteristics of democratic pragmatism (Dryzek 2005; Murphy 2017) in that they promote sustainable development through cooperation, equality, and collective problem-solving within the commons. Like elsewhere in the world, changing weather patterns have affected the livelihoods of farming communities in Kenya. For communities practicing pastoralism and living a nomadic lifestyle, foraging for pasture and water for their livestock, the effects of changing weather patterns (longer droughts, intense heat waves, seasonal uncertainties, etc.) can be especially difficult to bear. While pastoralist communities have been living this nomadic lifestyle for centuries, and one may argue that they are used to travelling longer distances, other challenges such as conflict between communities arise as they look for food and water for their livestock. How then can these communities adapt and survive the threat of climate change?

This is the question that motivated the creation of Afriscout. Also know by the moniker "the shepherd's eye in the sky," Afriscout was developed by Project Concern International (PCI), a global development organization as a way to provide pastoral communities with maps that furnishes pastoralists with information on where to graze and the conditions of various grazing areas. This way, the pastoralists are more empowered in knowing where to go and to share their knowledge about these locations. According to AfriScout, the impact of this has been a 48% reduction of livestock loss. It is interesting to see how these indigenous communities of pastoralists, whose lifestyle is threatened by climate change, adopt digital technologies to not only protect their livelihood but also protect their nomadic cultural lifestyle. This is a significant development because for some communities, adapting to the effects of climate change has meant giving up their cultural lifestyles, adopting new ways of living. This echoes Adger et al.'s (2012) assertion that "climate change threatens cultural dimensions of lives and livelihoods that include the material and lived aspects of culture, identity, community cohesion and sense of place" (p. 112). An app like this therefore also begins to reveal how the effects of climate change are not only experienced physically but can also cause cultural shifts.

Another application is M-Farm that is mainly used to provide farmers with price information on various produce. This app empowers the farmers who more often than not are based in rural areas on the market value of their produce in markets in the cities. The significance of this is that farmers are empowered with information such that they will not be short-changed by middle men who buy the produce from the farmer at a low cost and sell it in the city at a higher price. This way the farmer is able to make more of a profit for their produce. Relatedly, Ujuzi Kilimo, an application that provides farmers with precision farming information such as the quality of soil, fertilizers, changing weather patterns, and so on, seeks to help farmers make "climate smart" choices to anticipate changes and adapt their farming strategies to mitigate losses.

While all of these apps are all centred on making information to the farmers and various communities more accessible, climate change adaptation is of course more complex. But providing this information is an important step. Indeed, having accurate information is vital for climate change adaptation, which is why these budding apps are beginning to play a larger role in the lives of Kenyan farmers. While one might argue that not everyone has access to these technologies or can afford to consistently use these apps, the communal nature of these communities enhances the affordance of these apps, as they share the knowledge with one another.

For instance, with AfriScout, while previously pastoralists would rely on a few individuals to go out and look for pasture and come inform the rest on where to go, through the app those individuals can now disseminate information more efficiently even if only a few pastoralists use it, as the community is networked to share information through mobile phones and orally with other group members.

This collective knowledge sharing dynamic underscores the crucial role that ICTs are increasingly playing in climate adaptation, revealing the importance of community networks but also the presence of more "virtual" actors. For instance, AfriScout is comprised of Project Concern International (PCI), a global development institution as earlier mentioned, in partnership with the United States Agency for International Development (USAID), Google while working together with community elders to design the app. Ujuzi Kilimo is comprised of local and global partners and stake holders. These partnerships demonstrate what Unwin (2009) identifies as the significant role that institutional actors can play in Information and Communication Technologies for Development (ICT4D) projects. More specifically, these partnerships also illustrate the growing "local" involvement of institutions in the fight against climate change, as they begin to reflect on how climate change is not just a threat to society, but also a threat to their goals as institutions. It is thus no surprise that a technology company like Google that may seem to have little relationship with Kenyan pastoralism can identify connections between the services they offer and a need they can meet in terms of promoting climate change adaptation. The involvement of companies such as Google and platforms such as AfriScout is not only informational to the communities using them, as they become repository of information on the landscape (e.g., local knowledge) that can be used to trace the longitudinal impact of climate change over time. Some of these digital platforms thus play a much bigger role as the information they provide eventually form climate science data sets. This prompts us to think about how "big data" can be used to assist communities in climate change adaptation.

These interconnections underscore the growing impact of South-South relations where smartphone companies in China, through manufacturing more affordable phones, are able to provide more access of technologies (Ayodele and Sotola 2014; Qiu 2010), such that communities that previously lacked information vital for their livelihoods are now able to access this information. Moreover, experiences with these apps also reveal some of the ways that people in Kenya are engaged in facilitating climate change adaptations, as apps such as M-Farm and Ujuzi Kilimo which were created and are run by Kenyans, providing a strong reminder that communities are quite capable of participating in their own "development." Indeed, in many ways these apps are the epitome of "citizens' media," in that when technology is put in the hands of the people they come up with their own material in ways that encapsulate processes of democratization and social change (Gerace and Lazaro 2006; Rodriguez 2001).

In more boiled-down terms, what the case of mobile phones and app adoption in Kenyan demonstrates is what Pezzullo and Cox (2017) define as the central role of environmental communication, which is to "alert, amplify and engage" citizens to ecological issues. With varying degrees of success, these apps have been able to play these roles, whether in the form of alerting farmers of prices in the market or engaging them through nimble and meaningful climate change adaptation strategies to improve their yield, or to amplify the need for more resources to address the effects of climate change.

Local to National Mobile Phone Activism: The Case of Cape Town's Water Crisis

From 2014 to 2017, the city of Cape Town experienced one of its worst droughts and ran gradually out of water (Booysen et al. 2019; Muller 2017). To respond, the City Council imposed a six-level water restrictions strategy to avoid declaring "Day Zero," the moment when the city taps run dry and water supplies to the suburbs are shut off. For Muller (2017, 2018) and the Lancet Planetary Health (2018), this scenario impacting more than four million people was a failure in governance of a city once recognized for its careful stewardship of natural resources and focus on sustainability, and one of the world's greenest cities. They argue that the city management shifted from a science-based and risk assessment approach to a more populist one, thus endangering Capetonians in multiple ways. Regardless of the underlying reasons, the water crisis brought Cape Town to the frontline of international media attention through BBC, CNN, *The Guardian*, NBC News, and *National Geographic* (Lardieri 2018; Mahr 2018; Mezzofiore 2018; Welch 2018).

In the build up to a waterless city, the threat of "Day Zero" spread nationally in the form of a countdown. News of this dire countdown reached Talita van der Heever, a resident of the East London, a city more than 1000 kilometres from Cape Town. As a new mother feeding an

18-month-old child with baby formula, van der Heever immediately connected this crisis to her own dependence on clean and accessible water.

In an interview granted to CNN, she shared that when she realized that many mothers would not be able to feed their babies, she took action by turning to her mobile phone (Mezzofiore 2018)—a decision, it turns out, that started a grassroots movement.

> I went to the bathroom, flushed my toilet and recorded it on my phone saying, "This sound of toilet flushing, and tab opening is the sound we take for granted in our everyday life, and in Cape Town this sound is going to disappear soon." (Mezzofiore 2018)

She sent the note via WhatsApp to 300 people she knew in East London. This bathroom-generated call to action was heard and within three hours people began to volunteer to donate water to the at-risk communities of Cape Town.

The message had a ripple effect in three stages. First it spread to residents of the city of East London first, then to people and organizations in other parts of the country, and finally beyond the borders of South Africa. The sound of a flushing toilet quickly became a social media meme, and van der Heever's WhatsApp account was soon flooded with messages from people and companies who wanted to donate water. Responses came from the cities of Durban, Johannesburg, Bloemfontein, and even outside the country's borders, from Namibia. van der Heever reported this moment to CNN in these terms: "My phone doesn't stop ringing. Every time someone sends me a message on WhatsApp, by the time I finish reading it it's already moved down another 100 messages" (Alexander 2019). In a matter of 24 hours, thousands of five-litre water jugs were filled up and sent to drop-off locations established throughout the country. Consequently, 30 tons of water were collected in the city of East London alone, while Durban reported collection of 60 tons and Johannesburg 90 tons (Alexander 2019).

According to van der Heever, the call to action she issued provoked a sense of community throughout the country with people volunteering to help as much as they could. As awareness grew and people became more engaged, a strategy evolved to decentralize the movement by appointing representatives and drop-off locations in other cities where water was channelled. "Then I had to decide who needed water urgently, and thought of vulnerable people like elders" (van der Heever, personal conversation, May 5, 2019). At first she used her refrigerator as a board to plan and map the processes and campaign she suddenly found herself leading. It became an online and offline project where the smartphone served as the central tool to give instructions and receive feedback. The offline aspect was when she had to map the processes and write down places that needed to be served as well as the resources available (van der Heever, personal conversation, May 5, 2019).

Another challenge was how to transport and distribute the water collected to the beneficiaries in Cape Town. Here the activist came up with another idea, this time informed by her experience and knowledge of the farming market in her region. She reached out to farmers and truck owners that journey back empty to Cape Town after bringing their produce to Durban, East London, and Johannesburg. The idea was to take advantage of the empty cargos and have these vehicles transport the water jugs as they returned to Cape Town. Reaction to the idea was strong, with farmers pledging to transport water back to Cape Town at no cost.

Local schools were involved in the campaign with students attaching messages of hope to each bottle that's been sent out to Cape Town. van der Heever dubbed the early stages of the campaign as the "Talita Koumi Water Drive." However, as the scope of the operation grew, is reached out to the national disaster-relief organization, Gift of the Givers, to help manage the campaign. She was hoping that the organization could assist with the logistic and help establish drop-off points across the South Africa. However, "they did not take me seriously until the story appeared on the news. That was when they called me back" (van der Heever, personal conversation, May 5, 2019). Eventually, Gift of the Givers ended up launching a formal social media campaign with the help of van der Heever and others. The social media campaign was called Water 4 Cape Town and strives to preserve clean water by providing educational training on water saving tips and raising funds to make this possible (http://water4capetown.org/about-us/).

When thinking about what happened the day that she shared "the flush" and the message's eventual evolution to grassroots movement and then to formal campaign, van der Heever, who considers herself an activist and humanitarian, emphasizes how critical the WhatsApp app was in making things happen. "The response generated was hopeful as it shows how a nation can stand and bring about a solution to a specific problem" (van der Heever, personal conversation, May 5, 2019). The broader lesson, she says, is how much can be done by just a few individuals to mobilize action and move people to confront the many problems the world is dealing with.

Conclusion

ICTs in sub-Saharan Africa are being employed to inform and mobilize citizens around a host of environmental issues. The cases outlined in this chapter provide examples of how citizens have been able to use ICT tools to create local community action networks, build partnerships, and increase environmental awareness and share knowledge to address water shortage, share adaptive farming strategies, and create a campaign around e-waste.

While these cases only provide a small glimpse of what citizens are doing with ICTs to tackle environmental problems, they nevertheless suggest some important things about what kind of environmental activism might be productive in the future. First, African environment activists are already working hard to alert local citizens to local problems. While this starts with the deceptively simple act of "naming things" (e.g., calling something "e-waste" or a "water crisis"), it is in many ways a political act in that African's themselves are assuming the agency to call others' attention to environmental conflicts that may be ignored by or even caused by those in power. Second, ICTs have demonstrated the capacity to move this act quickly from the naming of something to increasing the number of people who are aware of it as a problem worthy of their attention. Significantly this amplification of awareness is an important step towards collective action, but, as was seen in all of these cases, can also be the critical "place" within environmental activism where potential partners are identified (e.g., AfriScout, the Gift of the Givers, local school, truck drivers, and Greenpeace Africa). Indeed, these apps are able to bring together various local and global stakeholders who understand the role that they play in addressing a particular problem. Finally, ICTs have a tremendous capacity to engage and facilitate community participation, as members use the information from these apps to dialogue and plan with each other locally, regionally, or even globally. While these steps are not necessarily linear, it is through these information and knowledge sharing and partnership building processes that communities are able to identify and adapt to environmental crisis as they make better and faster choices to problem solve or protect their livelihood.

To be sure, the growing role of ICTs in sub-Saharan Africa is a reflection of the influence and role that technological growth and globalization have had in the region. While ICTs are quickly becoming essential tools in the struggle to involve the public in sound environmental stewardship, they are also part of modernization, development, and globalization discourses that have been and typically still are anchored to exploitative relationships. Thus to gain deeper insight into how ICTs may lead to more sustained and even radical forms of environmental citizenship, one would need to do a more comprehensive ethnographic study with the creators of ICT-generated environmental campaigns and the communities of actors they attract and depend on. This would need to involve looking more closely at political subjectivity and agency in terms of how environmental problems are defined (e.g., what they are and whose responsibility is it to address them) and how communities are positioned within the problem-solving process (e.g., as active agents of change or merely as recipients of information).

References

Adger, W. N., Barnett, J., Brown, K., Marshall, N., & O'Brien, K. (2012). Cultural dimensions of climate change impacts and adaptation. *Nature Climate Change, 3*, 112.

Aker, J. C. (2011). Dial "A" for agriculture: A review of information and communication technologies for agricultural extension in developing countries. *Agricultural Economics, 42*(6), 631–647.

Aker, J. C., & Mbiti, I. M. (2010). Mobile phones and economic development in Africa. *Journal of Economic Perspectives, 24*(3), 207–232.

Alexander, C. (2019, April 12). Cape Town's 'Day Zero' water crisis, one year later. *City Lab*. Retrieved from https://www.citylab.com/

Avle, S. (2011). Global flows, media and developing democracies: The Ghanaian case. *Journal of African Media Studies, 3*(1), 7–23.

Ayodele, T., & Sotola, O. (2014). China in Africa: An evaluation of Chinese investment. *Initiative for Public Policy Analysis*, 1–20.

Banda, F. (2009). China in the African mediascape: A critical injection. *Journal of African Media Studies, 1*(3), 343–361.

Blankson, I. A. (2005). Globalization, pluralistic media and cultural identity transformation in emerging African democracies. *International Journal of Communication, 15*(1 & 2), 131–146.

Blankson, I. A. (2007). Independent media and media pluralism in Africa: Opportunities and challenges of democratization and liberalization. In I. Blankson & P. Murphy (Eds.), *Negotiating democracy: Media transformations in emerging democracies* (pp. 15–34). Albany: SUNY Press.

Booysen, M. J., Visser, M., & Burger, R. (2019). Temporal case study of household behavioural response to Cape Town's "Day Zero" using smart meter data. *Water Research, 149*, 414–420.

Carvalho, A., van Wessel, M., & Maeseele, P. (2017). Communication practices and political engagement with climate change: A research agenda. *Environmental Communication, 11*(1), 122–135.

Chandler, D., & Reid, J. D. M. (2016). *The neoliberal subject: Resilience, adaptation and vulnerability*. London: Rowman & Littlefield International.

Costanza-Chock, S. (2012). Mic check! Media cultures and the occupy movement. *Social Movement Studies, 11*(3–4), 375–385. https://doi.org/10.1080/14742837.2012.710746.

Dryzek, J. (2005). *The politics of the Earth: Environmental discourses* (2nd ed.). Oxford: Oxford University Press.

Edwards, S., Johnson, S., & Weil, D. N. (Eds.). (2016). *African successes, volume III: Modernization and development* (Vol. 3). Chicago: University of Chicago Press.

Eko, L. (2003). Globalization and mass media in Africa. In L. Artz & Y. R. Kamalipour (Eds.), *The globalization of corporate media hegemony* (pp. 195–212). London: Routledge.

Emeagwali, G. (2011). The neo-liberal agenda and the IMF/World Bank structural adjustment programs with reference to Africa. In D. Kappor (Ed.), *Critical perspectives on neoliberal globalization, development and education in Africa and Asia* (pp. 3–13). Rotterdam: Sense Publishers.

Garliardone, I., Kalemera, A., Kogen, L., Nalwonga, L., Stremlau, N., & Wairagala, W. (2015). In search of local knowledge on ICTs in Africa. *Stability: International Journal of Security & Development, 4*(1), 35, 1–15.

Gerace, F., & Lazaro, H. (2006). *Participation and communication. Communication for social change anthology: Historical and contemporary readings*. Princeton: CFSC Consortium.

Grätz, T. (2011). Contemporary African mediascapes: New actors, genres and communication spaces. *Journal of African Media Studies, 3*(2), 151–160. https://doi.org/10.1386/jams.3.2.151_7.

Grätz, T. (2013). Radio advertising and entrepreneurial conjunctions in Benin: Producers, styles and technologies. *Journal of African Cultural Studies, 25*(1), 42–56. https://doi.org/10.1080/13696815.2013.749779.

Greenpeace Africa. (2018). *Greenpeace Africa launches online campaign platform to strengthen environmental activism in Africa* 'press release'. Retrieved from https://www.greenpeace.org/africa/en/

Gumucio Dagron, A., & Tufte, T. (Eds.). (2006). *Communication for social change: Anthology: Historical and contemporary readings*. South Orange: CFSC Consortium.

Harrison, G. (2013). *Neoliberal Africa: The impact of global social engineering*. London: Zed Books Ltd.

King, K. (2013). China's aid and soft power in Africa. *The case of education and training*. Rochester: James Currey.

Kleine, D., & Unwin, T. (2009). Technological revolution, evolution and new dependencies: What's new about ICT4D? *Third World Quarterly, 30*(5), 1045–1067.

Krings, M., & Okome, O. (Eds.). (2013). *Global Nollywood: The transnational dimension of an African video film industry*. Bloomington: Indiana University Press.

Lardieri, A. (2018, January 31). Social media campaign collects water as Cape Town nears Day Zero. *U.S. News*. Retrieved from https://www.usnews.com/

Lester, L., & Hutchins, B. (Eds.). (2013). *Environmental conflict and the media*. New York: Peter Lang.

Mahr, K. (2018, May 4). How Cape Town was saved from running out of water. *The Guardian*. Retrieved from https://www.theguardian.com/us

Manyika, J., Cabral, A., Moodley, L., Moraje, S., Yeboah-Amankwah, S., Chui, M., & Anthonyrajah, J. (2013). *Lions go digital: The Internet's transformative potential in Africa*. Retrieved from https://www.mckinsey.com/industries/high-tech/our-insights/lions-go-digital-the-internets-transformative-potential-in-africa

Mbembé, J. A. (2001). *On the postcolony*. Berkeley: University of California Press.

Mezzofiore, G. (2018, January 31). Social media steps in to ease Cape Town water crisis. *CNN*. Retrieved from https://www.cnn.com/2018/01/30/africa/cape-town-water-campaign-intl/index.html

Moser, S., & Dilling, L. (2007). Toward the social tipping point: Creating a climate for change. In S. Moser & L. Dilling (Eds.), *Creating a Climate for Change: Communicating Climate Change and Facilitating Social Change* (pp. 491–516). Cambridge: Cambridge University Press.

Mudhai, O. F., Tettey, W., & Banda, F. (Eds.). (2009). *African media and the digital public sphere* (1st ed.). New York: Palgrave Macmillan.

Muller, M. (2017). Understanding the origins of Cape Town's water crisis. *Civil Engineering= Siviele Ingenieurswese, 2017*(v25i5), 11–16.

Muller, M. (2018). Cape Town's drought: Don't blame climate change. *Nature, 559*, 174–176. https://doi.org/10.1038/d41586-018-05649-1.

Murphy, P. (2017). *The media commons: Globalization and environmental discourses*. Champaign: University of Illinois Press.

Murphy, P., & Tinga, T. M. (2019). Communication for development and social change and the challenge of climate change. *International Journal of Communication, 13*, 1252–1270.

Musa, M. (2011). Media flows, domination and discourse in Nigeria. *Journal of African Media Studies, 3*(3), 329–348.

Nyamnjoh, F. B. (2013). *Africa's media democracy and the politics of belonging*. Pretoria: Unisa Press.

Oduro-Frimpong, J. (2014). Sakawa rituals and cyberfraud in Ghanaian popular video movies. *African Studies Review, 57*(2), 131–144.

Ogan, C. L., Bashir, M., Camaj, L., Luo, Y., Gaddie, B., Pennington, R., & Salih, M. (2009). Development communication: The state of research in an era of ICTs and globalization. *International Communication Gazette, 71*(8), 655–670.

Okolloh, O. (2009). Ushahidi, or 'testimony': Web 2.0 tools for crowdsourcing crisis information. *Participatory Learning and Action, 59*(1), 65–70.

Onwumechili, C. (2007). Nigeria: Equivocating while opening the broadcast liberalization gates. In I. Blankson & P. Murphy (Eds.), *Negotiating democracy: Media transformations in emerging democracies* (pp. 15–34). Albany: SUNY Press.

Pezzullo, P. C., & Cox, R. (2017). *Environmental communication and the public sphere*. Thousand Oaks: Sage.

Qiu, J. L. (2010). Mobile phones, the bottom of the pyramid and working-class information society in China. *The Electronic Journal of Information Systems in Developing Countries, 44*(1), 1–14.

Rodriguez, C. (2001). *Fissures in the mediascape: An international study of citizens' media*. Cresskill: Hampton Press.

Sparks, C. (2009). South African media in transition. *Journal of African Media Studies, 1*(2), 195–220. https://doi.org/10.1386/jams.1.2.195/1.

Thalén, O. (2011). Ghanaian entertainment brokers: Urban change, and Afro-cosmopolitanism, with neo-liberal reform. *Journal of African Media Studies, 3*(2), 227–240.

The Lancet Planetary Health. (2018). Water crisis in Cape Town: A failure in governance. *The Lancet Planetary Health, 2*(3), e95.

Tinga, T. (2019, May 1). *From Afro-pessimism to Africa rising: Anglo-American and Afro representations of the African continent*. Unpublished dissertation. Media and Communication, Temple University.

Unwin, P. T. H. (2009). *ICT4D: Information and communication technology for development*. Cambridge: Cambridge University Press.

VUMA.Earth. (2019). *People power local campaigns*. Retrieved from https://www.vuma.earth/

Wasserman, H. (2011). *Popular media, democracy and development in Africa*. New York: Routledge.

Wasserman, H. (2018). *Media, geopolitics, and power: A view from the global south*. Champaign: University of Illinois Press.

Welch, C. (2018, March 5). Why Cape Town is running out of water, and who's next. *National Geographic*. Retrieved from https://www.nationalgeographic.com/

Old Power Struggles and New Media Work: Indigenous Peoples' Striving for Justice in Contemporary Brazil

Camila Emboava Lopes and Annika Egan Sjölander

In August 2019, media outlets all over the world published alarming reports about devastating fires in the Amazon rainforest. Images of burning rainforest spread quickly, shared on social media platforms and throughout online networks. *The New York Times* reported that the hashtag #PrayForAmazonas was the trending topic in the world on Twitter (Londoño et al. 2019). World leaders around the globe also directed their attention to the Amazon. As the world's largest rainforest, its capacity to absorb carbon is widely understood to be crucial to our ability to tackle climate change. The international outcry for preservation of the rainforests and immediate action to extinguish the fires was not, however, welcomed by Brazil's President Jair Bolsonaro. Instead, he accused foreign leaders for 'colonialist' manners and disrespect of Brazil's sovereignty. Bolsonaro also criticized international 'deceitful' media for being

C. Emboava Lopes (✉) • A. Egan Sjölander
Department of Culture and Media Studies, Umeå University, Umeå, Sweden
e-mail: camila.emboava.lopes@umu.se; annika.egan.sjolander@umu.se

J. Díaz-Pont et al. (eds.), *The Local and the Digital in Environmental Communication*, Global Transformations in Media and Communication Research - A Palgrave and IAMCR Series,
https://doi.org/10.1007/978-3-030-37330-6_8

sensationalist and hyping the Amazon fire problem during his address to the UN General Assembly. In the same speech, he reiterated the government's plans to decrease the size of protected Indigenous land in the country (Philips 2019a). Bolsonaro's controversial approach was widely condemned by Brazil's Indigenous leaders. In their news coverage of the story *The Guardian* (Philips 2019b) referred to an open letter signed by 16 Indigenous leaders (Kalapalo et al. 2019), published on Twitter, shortly before Bolsonaro gave his UN talk.

Brazil is the largest country in Latin America and covers 47 per cent of the South American territory and has an estimated population of 210 million (IBGE 2019). The national survey of 2010 reports an Indigenous population of over 800,000 people, differentiated in 283 ethnic groups speaking 180 different languages (IBGE 2019). These different groups are spread over an enormous territory and live in very varied types of landscapes and ecosystems. Indigenous people in Brazil inhabit all states and live in both rural and urban areas. Some are living in small reservations, while others work or study in big cities. Demarcated traditional land constitutes homes for some, whilst other Indigenous groups live by the roads or in camps near land that are in judicial dispute. Violent confrontations with non-Indigenous landowners, particularly farmers are frequent in these places (United Nations 2016; Ochigame and Holston 2016). The tension has also increased in recent years.

A report regarding 2017, authored by the Indigenous Missionary Council (CIMI 2018), documented 20 cases of conflicts over territorial rights between Indigenous groups and non-Indigenous landowners, in addition to 96 instances of invasions, illegal exploitation of natural resources and damages to properties. Concerning violence against Indigenous individuals, they reported 110 deaths by assault, 27 cases of attempted murder and 14 cases of death threats. There were also 128 cases of suicide reported among Indigenous people in Brazil that year. The country is also extremely unsafe for environmental activists. A report from Global Witness (2018) shows that Latin America is again the most unsafe region for environmentalists. In 2017, Brazil recorded the most killings of any country with as many as 57 people killed. Considering the different scopes of these reports, it is not possible to know if any of the cases overlap. However, the high numbers show the turbulent context wherein which Indigenous people and environmentalists in Brazil live today.

Castilho (2012) investigated the so-called declaração patrimonial (declaration of assets) presented to the Electoral Justice in 2006, 2008 and 2010. His book discloses that the elected politicians own about 4.4 million hectares of land together, and this without including land owned by municipal politicians. In 2010, these candidates received around R$50 million (equals €11 million or US$12.1 million in today's worth) from companies linked to agribusinesses. In contrast to these figures, there is only one Indigenous representative in the Brazilian federal government in 2019. Historically and today, Brazil has been constructed and governed by an elite of landowners. The social structure of the nation has also developed with big properties as its base. Cavalcante (2013) claims that Brazil's systems are configured to maintain, what Quijano (2000) calls a 'coloniality of power', with a political system that benefits the interests of landowners.

In this chapter, we analyse different forms of resistance articulated to counteract what can be seen as a hegemonic view that legitimates power abuse against Indigenous people in Brazil. We focus the empirical study on what Norman Fairclough (1992, 1995) calls communicative events and have also structured the discourse analysis in line with his three-dimensional model. We in turn focus on text, the discursive practice including production and consumption of these text(s), and the wider sociocultural practice or context of relevance at work in the studied discourse(s). The chosen communicative events are two big protests regarding land use, organized by Indigenous people's movement in opposition to the Bolsonaro regime. These protests are, in many respects, at the centre of the current political turmoil in Brazil and pivotal moments of discursive struggles.

Our purpose is to outline how Indigenous people and, more precisely, media activists, construct their contemporary indigeneity and how they, in their own terms, articulate their identities on social media platforms, Facebook in particular. Ideas of land(-use) and nature/culture are very much part of these articulations and of special interest here. We also reflect upon the digital context and how its specific means of communication affects the dynamics between the local, national and global levels in the studied discourse since power relations are becoming more complex compared to the times before the wide introduction and use of Internet, including digital and social media (Lindgren 2017).

To give some context to the analysed communicative events, we start with a brief description of the current political climate in Brazil since President Bolsonaro took office, with a special focus on Indigenous peoples and their struggle for land. We then take a look at the importance of media and review some of the key literature from this rich field of studies.

Reflexivity, research ethics, methodological approach and the empirical material are then presented, followed by the discourse analysis of the two communicative events that we base our empirical study on. First we study all texts regarding the protests published on social media platforms by Indigenous media activists, then we discuss the discursive practice that forms this production and consumption of texts, followed by an analysis of the wider sociocultural practices that enable this particular discourse to emerge. We end the chapter with some concluding remarks.

Brazil's Indigenous Peoples, Land and Conflict

Year 2019 started in a very challenging way for the Indigenous rights movements in Brazil. On January 1, the right-wing nationalist President Jair Bolsonaro took office and immediately signed the Provisional Measure No. 870, transferring the Indigenous affairs agency called Fundação Nacional do Índio (FUNAI) from the Ministry of Justice to the newly created Ministry of Women, Family and Human Rights. In practical terms this excluded FUNAI from its role of demarcating Indigenous land and its environmental licensing task in Indigenous lands. These responsibilities were instead transferred to the Ministry of Agriculture, Livestock and Supply, a body which is basically controlled by a group of congress representatives who have pro-agribusiness interests as their main political agenda (De Alcântara 2019; Ochigame and Holston 2016).

This decision taken by the newly installed Brazilian president was the first step of many executed by the federal government enforcing a regime with a clear anti-Indigenous land rights agenda. Attacks on Indigenous rights were one of the central rhetoric arguments in Bolsonaro's electoral campaign, evident in the following statement:

> *The indian [indio (pt)] wants to become part of the society. I was played a dirty trick by some sectors of the media. I will repeat it here. The indians want electricity, physicians, dentists, Internet, and they want to play soccer. They want what we want. [...] Here in Brazil, some people advocate the idea to keep the indians in reserves as if they were animals in a zoo. I do not want that. I want to treat the Indians as human beings and citizens.* (De Alcântara 2019)

In response to the president, the organization Indigenous Peoples Articulation in Brazil (APIB) released a detailed letter:

> *We do not accept being treated as inferior beings, as your Excellency's statements seem to suggest. We are only different, and it is the Federal Government's obligation, according to the Constitution, to respect our 'social organization, customs, languages, beliefs and traditions' (article 231 of the Constitution). Therefore, we repudiate your disparaging and limited view in considering us as animals of a zoo.* (De Alcântara 2019)

Bolsonaro's statement is unapologetically assimilationist/integrationist. The president has repeatedly called into question the lawfulness of the current legislation which recognizes Indigenous people and their rights, while also accusing Indigenous peoples of being an obstacle to the country's economic development. The new government's approach regarding environmental protection appears to be even more problematic. Bolsonaro and his members of cabinet are also climate change deniers, and their political agenda is clearly directed at trying to stop existing environmental protection programmes, not least in the Amazonas rainforest. The president's anti-environmental approach has been widely criticized by international legacy media, such as *The Guardian* (Kaiser 2019) and *The New York Times* (Londoño et al. 2019).

The current political turmoil in Brazil has placed Indigenous social movements at the very forefront of the opposition against the nation's federal government. President Bolsonaro is a retired military officer and a member of the Social Liberal Party (PSL). During his two decades as a congressman, he has been mainly praising the armed forces and the former military dictatorship. His presidential campaign has leveraged social media platforms. Bolsonaro did not show up at public debates and generally avoided interviews with journalists, especially after he was physically attacked and seriously injured during one such public event. Throughout the election campaign (and after) Bolsonaro has received great and precious support from the popular and steady growing Neo-Pentecostal churches and leaders (Anderson 2019).

Direct presidential elections in Brazil reinaugurated democracy in 1989, including the creation of the nation's current constitution. This was written after the country had lived through the longest continuing military dictatorship of its history (between 1964 and 1985). This military regime culminated in human rights violations, restrictions of freedoms, exiles, tortures and murders (Napolitano 2018). Prior periods of (almost) democracy were intercalated with different forms of authoritarian governments—this after the country had been proclaimed a republic by a military movement in 1889. In 1822, Brazil gained independence from Portugal and the Portuguese, which had ruled the nation as a colony since 1500.

Indigenous Peoples and the Importance of Media

The importance of Indigenous-owned media in Brazil cannot be detached from the extremely unsafe, hostile, even violent conditions, Indigenous people experience in the country. The contemporary tensions are intertwined in many cases with old struggles for land. Indigenous people have been facing increasing threats during recent years in these times of political and economic turmoil in Brazil. A report published by the United Nations in September of 2016 asserted that Indigenous people in Brazil were at more risk in that year than in the final period of the military dictatorship in the 1980s. The document was produced by the Special Rapporteur on the rights of Indigenous people, Victoria Tauli-Corpuz, who concluded that Indigenous groups are facing increasing discrimination; stalled demarcation processes; violence, racism, killings and threats (United Nations 2016).

Like other Indigenous peoples in the world, Indigenous groups in Brazil are systematically misrepresented in history books and mass media (Mussi and Sousa 2013; Emboava Lopes 2016). Research reveals a consistent picture of the routine denigration of Indigenous peoples in dominant media outlets in Brazil and a frequent normalization of the interests of the landowners in the depiction of specific instances of conflict between landowners and Indigenous people (Cavalcante 2013). Stereotypes used, rooted in the complex socio-history of the country, often overlap with dehumanization processes, claims of primitiveness and discourses that constructs Indigenous individuals as incapable of thinking for themselves (Downing and Husband 2005; Alia 2010; Roosvall and Tegelberg 2015).

A significant amount of different media initiatives, both traditional and online, has emerged from Indigenous movements in recent years (Alia 2010). And as Foxwell-Norton, Forde and Meadows (2013, p. 138) argue, "The importance of Indigenous media for the community and their producers should not be under-estimated – but certainly the ability of Indigenous media to impact on 'mainstream' public sphere perceptions is limited." The use of newer, digital media technologies is often described as a source of hope, a powerful alternative, when it comes to media initiatives owned by Indigenous people themselves (Ochigame and Holston 2016). In Brazil, an increasing number of

Indigenous people use digital media technologies to create and share their narratives. The organization Indigenous Peoples Articulation in Brazil (APIB) and the multi-ethnic online radio station Rádio Yandê are significant examples of this movement of Indigenous peoples' increasing online presence. However, this visibility is always constrained by the logics of algorithms that are put in place by major social media platforms, such as Google, Facebook and Twitter (Ochigame and Holston 2016). Their filtering algorithms are built to determine both the order and the selection of items/data in a feed and are usually based on surveillance of user actions, like clicks and browsing histories. However, as Ochigame and Holston (2016) observe in their case study of social movements, filtering logics of platforms, and land disputes in Brazil between agrobusiness and 45,000 Guarani and Kaiowá people living in Mato Grosso do Sul, these influential algorithms are usually not disclosed, let alone explained, to the users. Ochigame and Holston (Ibid, p. 90) also contest the common claim that "filtering by popularity makes a platform democratic" since it is based on a very "limited concept of democracy". In their own empirical study of the filtering mechanisms used by, for example, Facebook, Twitter and YouTube, they showcase how these "hidden" algorithms used by global for-profit corporations tend to, for example, limit the distribution of conflictual messages. This 'filter of dissent' logic of course has an impact on the visibility and wider reach of the studied social movements' struggles for equality and justice. The critique of hegemonic views that Indigenous people publish on platforms does not tend to generate many likes or clicks, and therefore is not understood as 'popular' or ranked high in social media feeds more generally. And this even though land use and its problematic past ever since the colonialization began in the beginning of the sixteenth century is a pivotal part of an age-old conflict in Brazil.

Teun van Dijk's (2005) research about race and discourse in Latin America suggests that racism in this region usually has all aspects of European ("white") racism, however with much more complex ethnic and social bases. He further claims that racism in Latin America is a system of domination and power abuse in which 'racial', 'ethnic' and 'class' dimensions come together, one could say intersects, in many forms of inequality. At the centre is also, van Dijk argues, paradoxically the complete denial of its existence in Brazil and the rest of Latin America. Yet the vast differences in wealth and power are social problems rooted in racial inequality. People from African descent and especially Indigenous people are rarely in leadership positions in (mass) media productions, let alone part of any elites that control or own media in Brazil. Benites (2014) pointed out the tendency in Brazilian news reporting of relating Indigenous issues with negative themes such as violence and suicide, including the preference for landowners' perspectives and the image of the Indigenous as drunk, violent beings who attack each other and farmers.

During the 1980s, Indigenous rights movements experienced moments of significant revitalization. They expanded and strengthened their connections around the globe. The foundation of the Working Group on Indigenous Populations (WGIP) in 1982 was one important event that eventually came to publish the first United Nations Declaration on the Rights of Indigenous People (in 2007). The publication by the International Labour Organization called Convention 169, concerning Indigenous and Tribal People in Independent Countries, was also important. The discussion on the vital role of media for Indigenous movements developed as well in these forums, declarations and conventions. The relevance of media for Indigenous claims and Indigenous peoples participation in democracies is indicated in Art. 16 of the United Nations Declaration on the Rights of Indigenous People, stating, "Indigenous peoples have the right to establish their own media in their own languages and to have access to all forms of non-indigenous media without discrimination" (United Nations 2007, p. 14).

Similar calls for much more 'effective access to media' worldwide are still made by media and communication scholars today. They make the worrying observation that despite all hopes of new (read digital) technology improving democracy etcetera, media infrastructures and access have spread unevenly across the globe and reinforce inequalities between people and regions. In their contribution to the work of the International Panel on Social Progress (https://www.ipsp.org/), Couldry et al. (2018b, p. 180) highlight the need for all individuals and communities, Indigenous and others, to be able to "use media infrastructures to produce content, access information and knowledge, and actively participate in the realms of politics, culture and governance".

However, significant amount of media initiatives has emerged from the Indigenous movements in the last couple of decades. Research connected to these attempts have been carried out across the world, often aimed at specific peoples, initiatives or areas (cf. Foxwell-Norton et al. 2013; Roosvall and Tegelberg 2018). Rodríguez and El Gazi's (2007) work on Indigenous peoples' radio stations in Colombia shows the diversity within the field and how experiences vary between different communities. Indigenous media initiatives can also be related to Levy's (2018) broader concept 'peripheral media', which refers to media produced outside of mainstream society. These types of media, according to Levy's observations, have also been active and produced its own content, challenging their surroundings, not least in Brazil. Custódio (2018) has also found that social media and local media collectives, for example in Rio de Janeiro's favelas, have enabled Indigenous collectives and other marginalized communities to make their own voices heard. Valerie Alia (2010) introduced the term 'New Media Nation' in 2004 in order to describe the initiatives that emerged from the international movement of Indigenous peoples and gave it an international perspective to indicate its relevance. She asserts that "the New Media Nation is a Fourth World movement that is engaged with removing ethnic and national borders and placing pan indigeneity at the center" (Ibid, p. 15).

Reflexivity, Research Ethics and Methodological Approach

As two CDA-inspired and non-Indigenous researchers working with Indigenous issues, it is of utmost importance to reflect upon, critically review and problematize our own subject positions (Foxwell-Norton et al. 2013; Jørgensen and Philips 1999). One of us is from Brazil and has studied indigenous journalism in Sápmi, more precisely Kautokeino in Norway, including discriminatory mass media representation of Brazil's Indigenous groups (Emboava Lopes 2016). The other author is a Swedish discourse analytical scholar brought up in the northern part of the country that learned a lot from being seen as a representative for 'the majority society' when interviewing Indigenous Sami people for her thesis work about nuclear waste management (Sjölander 2004). Lawrence and Raitio (2016, p. 132) make the potent claim that "the choice between objective/neutral researchers on the one hand, and subjective/engaged ones on the other, is a false one. The choice rather, lies between researchers who acknowledge and critically reflect on their own role in (de)colonizing the (Nordic) academia, and those who do not". Decolonization is a pivotal concept within Indigenous research methodologies (Drugge 2016). It is an acknowledgement and a response to the long history of relations between academic research and colonization strategies. Decolonization is a process and can be understood in many ways. Here, we intend to inform our analysis with what Indigenous scholars from Brazil have written about land and rights and, in this sense, at least start to decolonize our perspectives when studying Indigenous peoples' own discursive constructions about themselves and their movements. The combination with a CDA approach works well as it seeks 'open' and interdisciplinary work, is focused on social problems and aims for 'bottom-up' perspectives (Fairclough 2003; Fairclough and Wodak 1997). Luke (2002) sees CDA as a text analytical approach that is explicitly normative or political, and van Dijk (2001), alongside with many others, argues that CDA aims to disclose how power abuse is enacted, reproduced and resisted in text. Chilisa (2011) outlines Indigenous research methodologies along a continuum: least indigenized, integrative, predominantly Indigenous and third space. We situate this chapter in the 'integrative' position, since it is informed by both non-Indigenous and Indigenous literature.

The 'order of discourse' to speak with Michel Foucault (1971/1993), where Indigenous peoples are consistently framed by mainstream media in stereotype and negative terms, has concrete implications. Discourses, this language-in-use, indeed have social implications. As Fairclough (1992, p. 64) states, "discourse is a practice not just of representing the world, but signifying the world, constituting and constructing the world in meaning." Discourses are both constitutive of the world and constituted by them and forms our understanding of things, ourselves and others (Fairclough 2017; Egan Sjölander 2011; Jørgensen and Philips 1999). Hence, it really matters what they communicate and contain.

Our analysis is also inspired by James Clifford's (2013) understanding of 'indigeneity' as a dynamic term, a work in progress. Clifford (2013, pp. 36–41) proposes three analytic terms to understand indigeneity in contemporary times, namely articulation, performance and translation. All these work as metaphors that indicate processes and complement each other. Articulation is a broad range of connections and disconnections and capture negotiations of alliances. Performance is the process through which Indigenous groups make themselves legible to other audiences, and refers to how they choose to represent their own identities, what to reveal or conceal. Translation is the dialogic process of communication and mediation of indigeneity with the wider society and refers to cultural ideas and practices that are transformed and reinvented during the interaction with others.

When researching Indigenous-owned media, it is necessary to briefly discuss the complex, and by no means self-evident, term 'Indigenous'. In this chapter, we use it since it is how Indigenous movements in Brazil 'movimento indígena' identify themselves to refer collectively to a broad range of peoples. Following literature on Indigenous research methodologies in English language, we write the term with a capital, which is connected to a pan-national identity in order to differentiate it from the adjective 'indigenous'. Downing and Husband (2005, p. 122) claim that it is crucial to understand Indigenous people's experience—the historical colonial relations that included usurpation of Indigenous rights and lands, as well as imposed "alien definitions of identity upon them". The stereotyped portrayal of Indigenous peoples recalls that, historically, descriptions of native groups were made by colonial explorers, settlers and scientists. The emphasis in these descriptions was often on primitiveness. The United Nations' Permanent Forum on Indigenous Issues has adopted an understanding of Indigenous people that can better be described as a 'non-definition', which we sympathize with. According to it, the most useful approach is to identify Indigenous peoples rather than defining them. This process favours self-identification at an individual level and acceptance by the community (United Nations 2018). In this project, we have worked with the ethical framework (Version 2.0) that the Association of Internet Research (AoIR) recommends, which advocates a process approach and attention to specific needs of each project (Markham and Buchanan 2012).

The analysis of the communicative events that we base our empirical study on is structured according to Fairclough's model with text, discursive practice and sociocultural practice as main analytical dimensions. The social media posts about the protest published from January to April 2019 by APIB (on Facebook) and Rádio Yandê (on Facebook, Instagram and Twitter) form our empirical corpus of texts. No software or computational method for scraping data from the platforms were used. Instead, we manually monitored these platforms and collected all posts about the protests, read through all texts and then made a selection and closer reading of the ones that show how media activists construct their own identity and contemporary indigeneity.

Articulated ideas about land(-use) and nature/culture were of particular interest. The analysis of these texts also, to some degree, considers the interaction among users, more specifically reactions and shares of these posts since they are part of the discursive practice. The numbers about post reactions were calculated on July 1, 2019.

The two significant protests about land rights that we focus the analysis on were organized by Indigenous movements and held during the first four months of 2019. The first protest took place on January 31, when the Articulation for Indigenous People in Brazil (APIB) together with local Indigenous associations, organized marches with the slogan #RedJanuary in different areas of Brazil. The aim of the protests, according to the media activists, was to call attention to the growing threats to Indigenous rights and territories. The NGO Survival International simultaneously organized protests in front of Brazilian embassies in North America and Europe. In total, protests took place in 71 different locations (cf. Fig. 1).

The second protest happened during the annual Free Land Camp (FLC) protest that was held between 24 and 26 of April, with APIB as main organizer. Around 4000 Indigenous people from all over Brazil gathered in the federal capital Brasilia and camped near the parliament. Marches, public acts, meetings with political authorities, assemblies and debates were carried out during these days. In Europe, the international high-profiled environmental campaigner Greenpeace organized support protests in front of several Brazilian embassies. The FLC protests started already in 2004 in opposition to celebrations of Brazil's 'discovery' and have since been developed against former Labour Party President Lula's government and how they have handled the demarcation of Indigenous lands (Guerra and Valiente 2014).

APIB was founded in April 2005 during the second FLC protest. Since then it has become the organization that coordinates, strengthens and supports Indigenous associations from different areas of Brazil. APIB also organizes national meetings, and the annual FLC is the biggest one. They also attend international Indigenous meetings. When it comes to their external communication, APIB uses a website (APIB 2019a) and a WordPress blog (APIB 2019c). Their Facebook page is followed by nearly 55,000 people (APIB 2019b) and their Twitter account has 3525 followers (APIB 2019d).

The content about the marches and protests were mainly created by APIB and then echoed by a number of supporters, activists and media organizations, including the pioneer station Rádio Yandê (Rádio Yandê 2019a). This first web-based Indigenous radio station in Brazil was created in 2013 by a collective of young Indigenous individuals from different ethnic groups. The station operates from its own website and makes use of different popular social media platforms. They use Facebook (45,664 followers), Twitter (3102 followers), Instagram (16,402

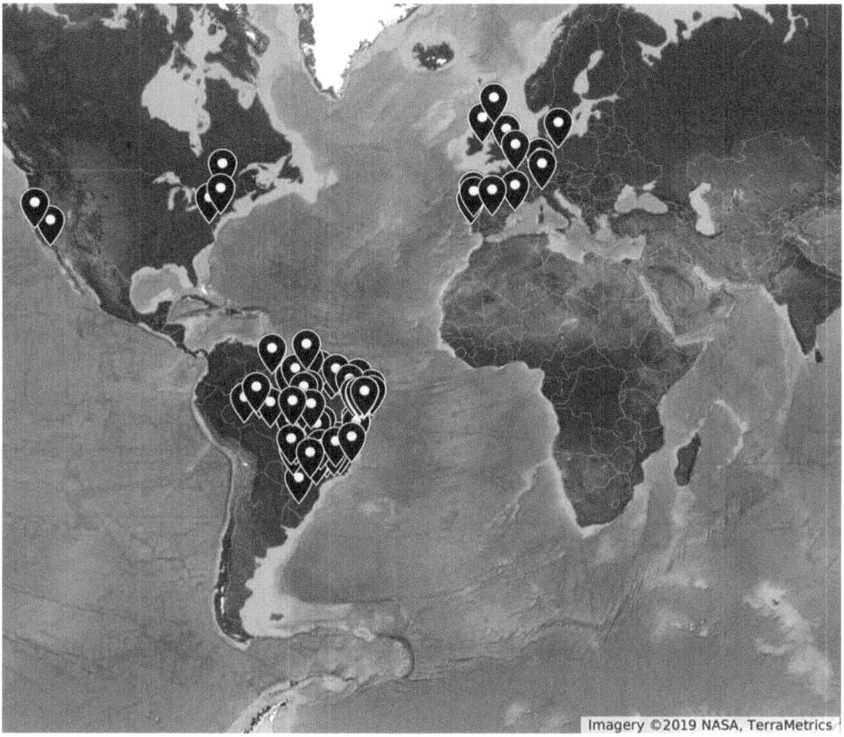

Fig. 1 Map created with Google My Maps shows #Red January events organized in 71 different areas. The marches, protests or performances took place at different times and formats. (Available from: https://bit.ly/2kDeFpf)

followers), YouTube (622 followers) and SoundCloud (319 followers) (Rádio Yandê 2019b, 2019c, 2019d, 2019e, 2019f). All content broadcasted by Rádio Yandê is produced by Indigenous peoples from all over the world, mainly from Brazil. However, most of it is not produced by the radio's staff. Some of the songs, and even interviews, are linked to a video on YouTube (that opens on a new page). Rádio Yandê uses social media mainly to share, retweet, repost, like and tag other users. These interactions and the hyperlinks to YouTube strengthen the sense of being a network: a set of connections between Indigenous communicators, musicians, artists, vloggers and their audiences. The radio station appears to operate with a very scarce and unstable budget. It is financed through donations (crowd sourcing) and different temporary projects and events about art, media and music that the founders and their collaborators manage to receive.

Because of the importance of Rádio Yandê as a self-owned Indigenous media organization and of APIB as the key organizer behind the studied communicative events, we have looked at the content (texts) published by Rádio Yandê on Facebook, Instagram and Twitter, as well as APIB's posts on their Facebook page from January to April 2019. The choice of analysing APIB's posts on Facebook, rather than their tweets, is due to the popularity and the large number of Facebook users in Brazil (Newman et al. 2019). The message app WhatsApp is also very popular among Brazilian citizens and is used by both APIB and Rádio Yandê. However, since the messages that are sent via this app are not public, but directed to specific individuals (private in a sense), we have not included them in our empirical data collection.

Two Communicative Events: A Text Analysis

The Red January Protests

APIB and Rádio Yandê's first Facebook posts 2019 are directed to the new President Bolsonaro (PSL) immediately after he transferred the authority to demarcate Indigenous lands from the Indigenous affairs agency FUNAI to the agriculture ministry. APIB published the following statement:

> *For the right to exist! For a pluri-ethnic and multicultural Brazil!*
>
> *(…) Bolsonaro and the colonels of the ruralist caucus know that to put more land on the market, they will need to make the demarcation of Indigenous lands impossible (…). But they also know that the world tends to new ways of production and consumption, and that we will not hesitate to denounce this government and agribusiness in the four corners of the world, demanding the adoption and respect to the social and environmental guidelines that are necessary to the accomplishment of our constitutional rights. (…)*
>
> *We Indigenous peoples, respecting our ancestors and compromised with the future generations, are willing to defend our livelihoods, our identity and territories with our own life. And we invite the Brazilian society to fight with us for a more fair and solidary country and for our right to exist.*
>
> *For that, the Indigenous Peoples Articulation in Brazil – APIB recommends that each State organizes the start of a judicial action requiring the nullity of the actions of the president Jair Messias Bolsonaro that practically*

destroys all the Brazilian indigenist policy. Demarcation now! (APIB, January 2, 2019) (Original in Portuguese; translation by the authors)

On Facebook on January 3, Rádio Yandê shared a letter formulated by Indigenous leaders from the north of Brazil:

> *We have already been decimated, under tutelage and victims of integrationist policies of governments and the Brazilian National State, so we publicly assert that we do not accept any integration policy, tutelage policy and do not want to be decimated by the means of new government actions and the Brazilian National State. This country called Brazil owes us an unpayable value, Mr. President, for all that has already been done against our peoples. Indigenous lands play a very important role in maintaining the richness of biodiversity, air purification, environmental balance and the very survival of the population of Brazil and the world. /…/*
>
> *We have the full civilian capacity to think, discuss the paths for Indigenous people according to our rights, which are guaranteed in Articles 231 and 232 of the Federal Constitution, Convention 169 of the International Labour Organization (ILO) and the UN Declaration on Indigenous Peoples.* (Rádio Yandê, January 3, 2019) (Original in Portuguese; translation by the authors)

APIB's post was shared over a thousand times and received 898 reactions (mainly 'likes') on the platform. Rádio Yandê's post was shared 12 times on Facebook and had received 32 likes. Letters/posts/texts delineate central points in understanding how Indigenous peoples of Brazil construct their indigeneity, including how their struggle for justice and land in opposition to the new government is performed. First of all, they stress that the construction of this country called Brazil has a history of Indigenous exploitation; secondly, they express an understanding of the importance of Indigenous lands for the global environment; and thirdly, they remind their audience of their constitutional rights and signed international conventions.

Altogether, there are 69 posts about the Red January protests published on APIB's Facebook page. The first posts are banners calling for mobilizations, then comes banners spreading information about the location and times of events, and finally there are a stream of pictures and videos from actions that took place. While scrolling through those protest accounts, the diversity of events becomes visible. Rádio Yandê's posts follow the same pattern: calls for mobilization, information about events and pictures from the marches. Many of their shared campaign posts (banners and videos) were made by APIB. During the protests, APIB and Rádio Yandê posted pictures sent in by protesters, evincing the social media strategies of interactivity and first-person narratives (cf. Lindgren and Cocq 2017).

The first Red January campaign banners state in capital letters; "Soon our hands will be united in a common struggle" which is followed by: "Organize your territory. Promote circle talks, debates, public acts, judicial, collective denounces, mobilizations and wait for the call". In the days shortly before the protests, the banners were changed to "Indigenous blood/Not another drop", written in red characters. The used hashtags during the protests were #JaneiroVermelho (#RedJanuary) and #DemarcacaoJa (#DemarcateNow). Another slogan of APIB was "The struggle is persistent". The mentioning of Indigenous blood can be seen as a challenge to the normative national narrative of Brazil as a benign ethnic melting pot (Ribeiro 1995). Indigenous leaders have been stating that in Brazil everyone has Indigenous blood indeed, some inside their veins and others on their hands. The idea of 'hands united' can be read as dialoguing with not only diverse Indigenous communities, but also other groups that oppose the new government. After the election results, an image of a rose and two hands holding each other with the phrase "Ninguém solta a mão de ninguém' (No-one lets anyone else's hand go) went viral since it was shared by many opposition groups (Pimentel 2018).

Considering the different platforms, Rádio Yandê mostly used Facebook to share information about the campaign. The most frequently shared post by both APIB and Rádio Yandê (1100 times in total) was a video on Facebook of young people mobilizing people to protest (made by APIB and Mídia Índia). It generated nearly 17,000 views and got 255 positive reactions and supportive comments. The most shared 'national' event post was APIB's post with pictures from the march in the city of São Paulo. It was shared 778 times, got 770 reactions (532 likes, 235 love and 3 wows) and 26 comments mainly consisting of support messages and tags. Its popularity is probably related to the presence of the famous Brazilian rapper, Criolo. The most shared 'international' event post (290 times) was of the action in London, which also had the presence of celebrity. Celebrities have come to play a more prominent role in today's environmental communication (Anderson 2011). However, such presence is a double-edge sword. It can draw attention to an issue, but at the same time be a distraction.

A shared GoogleMap posted by APIB, very similar to the one the #FridaysForFuture-movement use in their climate change communication, with information about the confirmed marches and related events in 71 different places, illustrates how the local and the global, through the use of digital tools for spatial visualization, becomes intertwined in this discourse. This post was viewed nearly 20,000 times and illustrates, as Salovaara (2016, p. 827) argues, that "digital mapping is by nature a dynamic, performative, and participatory practice".

The Red January protests did not gather huge actual crowds. However, as a communicative event it connected different distant locations to each other. These activities also worked as a prelude to the 'bigger' protests that happened in April. Similarly, Lindgren and Cocq (2017) have analysed information sharing, network building and support on Twitter in relation to Sámi anti-mining protests in 2013 in Sweden. They conclude that Sámi actors were at the centre of communication and their messages were spread out to a wider network. And this in turn indicates the relevance of social media for geographically distant groups and marginalized communities when they try to make their claims heard globally. Deem (2019) also observes in her case study of the No Dakota Access Pipeline-movement, or the #NoAOL-hashtag, that the affordances of social media work to unite Indigenous epistemologies with Western science-based environmentalism. The Red January protests however seem to have generated little interest from traditional mainstream media; at least any such cover is hardly mentioned by, or redistributed via, APIB or Rádio Yandê.

The Free Land Camp Protests

APIB's page on Facebook counts 216 posts about the Free Land Camp (consisting of banners, pictures, videos, live streams, texts and hyperlinks). Before the camp started, posts were asking for donations to cover protesters' travel and food costs, but also included responses to the federal government's decision to call up the National Force to protect Brasilia during the duration of the FLC protests. This move was interpreted by people from Indigenous movements as an attempt to intimidate the upcoming camp. All activities, from marches to cultural performances and official meetings with politicians, were shared on APIB's Facebook page during the camp. And because of this, it was also possible to follow the decision to relocate the camp to a more distant location from congress. This in order to avoid physical confrontations with the heavily armoured National Force.

The majority of APIB's posts portray a diverse group of people joining the camp. Perhaps a quick look can relate the feathers and traditional body painting to stereotyped non-Indigenous views of indigeneity. However, a more careful look reveals the diversity of ethnic groups in Brazil, reflected in specific way to dress and perform (Clifford 2013). Activities such as the women's meeting and the youth gathering also nuance the Indigenous movement by gender and age. This can be a positive step in avoiding stereotypes that represent Indigenous as homogeneous or 'frozen' in past times.

Rádio Yandê posted about the Free Land Camp on Facebook (11 times), on Instagram (6) and Twitter (9). The first posts are calls for donations and invitations to the camp. The third post is from mainstream media about the president's request for the presence of the National Force in Brasilia. On April 22, when Earth Day is celebrated, Rádio Yandê posted videos about it on Facebook and Instagram. The videos were clearly aimed at global articulations for their struggles, since there were Portuguese, English, Spanish and French versions. They included images of former editions of the Free Land Camp and a rap song that mixtures Portuguese and Guarani languages. The Guarani language belongs to the Tupi-Guarani linguistic family (Grünberg and Melià 2008). It is spoken by over 100,000 Guarani Indigenous peoples in Brazil, and it is one of the official languages in Paraguay, where it is spoken by around six million people. Below is one of the Facebook posts that highlight the struggle for justice and land:

> *Indigenous blood. In the veins, the fight for land. Free Land Camp. National meeting in defence of Indigenous people rights. April 24–26 Brasilia. Thousands of Indigenous leaders walk to Brasilia. Indigenous lands to preserve our culture, nature, air and water. Jair Bolsonaro launched a major offensive against Indigenous rights. We fight for climate justice, we fight for human rights. We are peaceful. We want justice. Justice! Support Indigenous peoples. Support the 2019 Free Land Camp of the Articulation Indigenous Peoples of Brazil.*

The campaign banners of the Free Land Camp show a fist holding an archery and an arm with veins that transform into a tree with big roots, which relates to the camps motto "Indigenous blood. In the veins, the struggle for land" (Fig. 2). The predominant colour of both the studied campaigns was red. Colours have also been central in the process of political polarization in recent years in Brazil. Anti–Labour Party movements

Fig. 2 Banner of the #Free Land Camp campaign. Arm and hand that hold the archery emerge from the soil. The veins are roots connected to the land. Red is the prevalent colour. (Available from: https://bit.ly/2kva1tx)

have been marching in the national flags' colours: yellow and green. The shirt of the Brazilian football team has become an icon of anti–Labour Party marches and then of pro-Bolsonaro marches. Nationalist right-wing movements have been repeating that 'Our flag will never be red' (Huffington Post 2019). Indigenous movements' response to this slogan is to state that Brazil's flag is red already from stains of Indigenous blood. During both communicative events some of the protesters also carried a Brazilian flag with stains of 'blood' to visualize their message.

The word 'justice' had a central role in the Free Land Camp. Some posts show that during the first evening march of the camp, protesters, for example, formed the word 'Justiça' with LED lights on the ground. This particular act had the support of Greenpeace. Beside more general claim for justice for Indigenous peoples, a concrete demand in these protests was that the Indigenous Affairs Agency FUNAI would be transferred back to the Ministry of Justice. And unlike Red January, the annual FLC protests received considerable media coverage, both nationally and internationally. Big international outlets such as BBC, Reuters, *The New York Times*, *The Guardian* and Al Jazeera reported about the protests. At the national level, several 'old' legacy media published FLC stories, including newer popular websites such as G1 (from Grupo Globo) and *Folha de São Paulo* (from Grupo Folha) (Alexa 2019). National versions of global media outlets, such as *El Pais* Brasil, *DW* Brasil and *Huffington Post* Brasil, also gave attention to protests.

Discursive Practices of Indigenous and Social Media

Alia (2010) describes the "New Media Nation" as an outlaw organization and claims that in many contexts Indigenous media relates somehow to warfare ideas. For instance, Maori videographer Dean Te Kupu Hapeta states that "Media is our non-violent way to wage war" (Alia 2010, p. 161). Alia observes that "in some places, 'outlaw' or 'guerrilla' media are a necessary part of survival" (2010, p. 109). This is closely related to Downing's (2001) discussion about radical alternative media, which he defines as generally small-scale media initiatives that voice alternative visions to hegemonic perspectives. "Radical media could easily be read in two different ways: as necessary to build counterhegemony but only truly powerful at times of political upsurge"(Downing 2001, p. 16).

Similarly, Indigenous media producers in Brazil many times describe their activities as a form of resistance or nonviolent war. For example, in a text published by the website globallandscapesforum.org, Indigenous media makers claim that they use camera lenses as weapons, which makes them stronger (Gutierrez 2019). In the case of the Red January and FLC protests, it was very clear that APIB and Rádio Yandê were using social media to mobilize resistance towards the current political regime and to counteract its hegemonic views. However, as Ochigame and Holston (2016) illustrates with their work on filtering algorithms, the struggle in social media for Brazil's Indigenous groups can be a very difficult one to win. Critical standpoints and dissent in general do not generate, for example, many likes on these platforms, which in turn have a negative impact on visibility.

President Bolsonaro has a very strong presence on social media platforms with 13 million followers on Instagram, around 11 million followers on Facebook and 4 million on Twitter (data from July 20, 2019). Brazil's president, like the US President Trump, also discredits mainstream media and prefers to make official announcements via his own social media platforms (cf. Boczkowski and Papacharissi 2018). Bolsonaro also produces weekly live streams for Facebook. In one of his live videos at the start of April 2019, the president states that Indigenous groups will use public tax money to pay for the Free Land Camp. On April 12, APIB denied this accusation in a video published on Facebook and claimed that it was not true since it is not even possible according to Brazilian legislation. They also stressed that FLC is financed by APIB's own funds and via donations and ended the video asking viewers for more support. APIB also shared a note denouncing the president's attempt to discredit their movement.

APIB's posts about the FLC protests show that articulations with politicians in opposition were part of the events. Visits and messages of support from politicians were welcomed during the protests. These articulations took place parallel to the Indigenous movements' own agenda to try to gain more seats within political parties. During the 2018 elections, the Socialist Party's (PSOL) Sonia Guajajara was the first Indigenous woman to run for vice president. During the same year, Joenia Wapichana managed to be elected as a Congress member via the Sustainability Network Party (REDE), and she also became the first Indigenous woman in Congress.

On Facebook APIB shared the Free Land Camp texts published by international and national mass media. Many of the quotes in this news reporting seem to be from the collective interview with the media that was organized during the first day, or from public speeches during the protests. The mainstream media coverage of FLC seems in general to harmonize well with the articulations posted by APIB and Rádio Yandê. A similar pattern of 'symbiosis' was also observed at another recent protest for Indigenous rights, the protest camp against the Dakota Access Pipeline in the United States known as NODAPL. And when it comes to the role of social media in their protests, Moore (2019, p. 180) asserts that "the tribe made clear that social media cannot only bypass traditional news but also work in tandem with it, even spurring traditional media outlets to cover news they normally would not".

Sociocultural Practices of Colonialization and Marginalization

Being colonized and marginalized is one of the common experiences of indigeneity across different groups, in Brazil and globally. Coulthard (2014) makes a powerful distinction between class exploitation and the colonial dispossession of Indigenous peoples. In Brazil, the concrete presence of Indigenous peoples in the land has constituted an ongoing challenge to the production of a coherent national narrative of Brazilian identity. The continuing denial of the (land) rights of the Indigenous population and the explicitly brutal treatment that they receive destabilize the myth of racelessness in the discourse. The claim by Indigenous peoples to the centrality of their historical distinctiveness, which is key in their self-identification as a "national Indigenous minority" (cf. Kymlicka 1995), perversely leads to them being positioned as a primitive, premodern anomaly in the body of Brazilian cohesion. In this study of Indigenous media activists' articulations of identities on social media platforms, we observe that the struggles for land rights and environmental protection are pivotal in this discourse. The colonization of Brazil was a cultural clash in many aspects, fundamentally between diverse forms of territorialization and the imposition of a specific form of territoriality by the colonizers. The relation to territory is essential to the foundational stories of Indigenous identity as is the continuing pervasive relevance of land to collective identities and cultural integrity. Gersem dos Santos, researcher from the Baniwa people, explains that the general Indigenous understanding of land is connected to the idea of a territory that is essential to the development of their traditions, social relations and cultural manifestations. In that sense, Indigenous peoples develop a close and deep relationship with the land—that is not only a physical space, but has cosmological meanings (Dos Santos 2006).

As an example, we can look at the umbilical relation of the Guarani and Kaiowá people with the land. This relation is expressed in the concept of 'tekoha', which is a term in the Guarani language dealing simultaneously with social group, geographic space and cultural system. In Guarani language, *teko* means way of being and living, and *ha* means place (Benites 2014). As geographic space it represents the area that has environmental conditions to develop the cultural system that defines their ways of life: material resources and spiritual elements (Oliveira and

Pereira 2009). This powerful association of land with spiritual and social identity is clearly present in other Indigenous peoples' epistemologies too and marks a potent source for politicization of space in relations between Indigenous peoples and settler societies. The cultural integrity of Indigenous rights movements in Brazil is locked into a nonnegotiable identification with their homeland: an identification which puts them in direct conflict with the interests of an elite of landowners and their partners.

The echoing of Indigenous voices on social media are articulating, performing and translating their identities in the struggle for justice on their own terms. These digital discourses can perhaps help to move away from the idea about a primitive 'indian' frozen in time that needs to be absorbed by the majority society. The official policy on indigeneity in Brazil until the 1970s was assimilationist: Indigenous peoples from many different areas were confined in reservations with the formal aim of gradually being educated and socially integrated as Brazilian citizens, while their lands were cleared up to be transformed into profitable farms.

However, the Constitution of 1988, promulgated when the country was leaving a military dictatorship, represented a great shift in the relations between the state and Indigenous peoples. It was the first time that Indigenous rights for land, language and culture were recognized (Amado 2014). The constitutional text established the right for Indigenous peoples to live according to their own worldviews and to have their traditional lands demarcated. In practical terms, the Constitution determined that Indigenous lands should be demarcated by FUNAI within five years. The demarcation processes are very complex however and confounded by judicial processes. More than 30 years later, most of these demarcations are not yet completed.

The search for global connections is a common feature that Indigenous peoples movement share with environmentalists. Lester (2016, p. 8), for example, observes that "campaigners are regularly 'speaking' to the distant, attempting to invoke a transnationalized public sphere", in her analysis of the debate over the Great Barrier Reef in Australia. Foxwell-Norton and Lester (2017, p. 575) have also showed that the communication strategy applied in the Fight for the Reef's campaign "is aimed at transcending national boundaries and bypassing traditional channels of power associated with media, government and industry".

Concluding Remarks

In the social media posts about the protests, media activists form APIB and Rádio Yandê construct their indigeneity and they make firm claims about (land) rights referring to their close relationship with land and nature. They also propose a radical decolonization of the meaning of (ownership of) land. The translation of their struggle and claims to the majority population was, at least in part, mediated by digital technology. We also noticed a pattern in the studied discourse which started with calling for support, then encouragement for mobilizations and alliances, then performance(s) of their identities and claims, and, finally, translating processes with a broader public, which in turn might be a way of gaining more support.

This pattern fits well with Clifford's (2013) conceptual tools of how to understand indigeneity in contemporary times. Articulations were made between local and national Indigenous peoples movements, between national and international NGOs, and with different politicians. Media activists' indigeneity was further performed during the marches and other forms of protests and communicated at social media platforms. Indigenous peoples movements' claims for the demarcation of their lands were also to some extent translated to the broader majority society of Brazil, including globally via, for example, international (mass) media outlets. In the dialogical process of translation, Indigenous peoples' demands for justice and land rights were also translated in terms of environmental protection and their long-standing connection with nature/land.

And finally, when it comes to the relationship between the local and the global in these communicative events, digital technology in the form of social media platforms turned out to play a vital role since their use strengthened connections between distant geographical places. However, the crucial role of online practices could not substitute the importance of grassroots mobilizations, marches and protests. The studied resistance by Indigenous groups also had concrete impact on Brazilian politics. Because at the end of May 2019, the Congress voted to cancel measure 870. That meant that they placed FUNAI back at the Ministry of Justice, including its authority to demarcate Indigenous lands. And even if Indigenous movements of course celebrate this latest development as a significant victory, APIB (2019) also fears that FUNAI may still be at risk because of budget cutbacks. Expressed in other words, old power struggles remain despite new media work in the striving for justice and land in contemporary Brazil.

References

Alexa. (2019). *Top sites in Brazil.* https://www.alexa.com/topsites/countries/BR. Accessed 5 July 2019.

Alia, V. (2010). *The new media nation: Indigenous peoples and global communication.* New York: Berghahn Books.

Amado, L. H. E. (2014). *Poké'exa ûti o território indígena como direito fundamental para o etnodesenvolvimento local.* MA thesis, Universidade Católica Dom Bosco, Campo Grande.

Anderson, A. (2011). Sources, media, and modes of climate change communication: The role of celebrities. *Wiley Interdisciplinary Reviews, 2*(4), 535–546.

Anderson, P. (2019). Bolsonaro's Brazil. *London Review of Books, 41*(3), 11–22. https://www.lrb.co.uk/v41/n03/perry-anderson/bolsonaros-brazil. Accessed 24 May 2019.

APIB. (2019a). *Articulação dos Povos Indígenas do Brasil homepage.* http://apib.info. Accessed 23 July 2019.

APIB. (2019b). *Articulação dos Povos Indígenas do Brasil official Facebook page.* https://www.facebook.com/apiboficial. Accessed 23 July 2019.

APIB. (2019c). *Weblog Mobilazação Nacional Indígena of Articulação dos Povos Indígenas do Brasil.* https://mobilizacaonacionalindigena.wordpress.com. Accessed 23 July 2019.

APIB. (2019d). *Articulação dos Povos Indígenas do Brasil official Twitter page.* http://www.twitter.com/apiboficial. Accessed 23 July 2019.

Benites, E. (2014). *Oguata Pyahu (Uma Nova Caminhada) no Processo de Desconstrução e Construção da Educação Escolar Indígena da Reserva Indígena Te'ýikue.* MA thesis, Universidade Católica Dom Bosco, Campo Grande.

Boczkowski, P. J., & Papacharissi, Z. (Eds.). (2018). *Trump and the media.* Cambridge, MA: MIT Press.

Castilho, A. L. (2012). *Partido da terra: como os políticos conquistam o território brasileiro.* São Paulo: Contexto.

Cavalcante, T. L. V. (2013). *Colonialismo, Território e Territorialidade: a luta pela terra dos guarani e kaiowa em mato grosso do sul.* Dissertation, Universidade Estadual Paulista, Assis.

Chilisa, B. (2011). *Indigenous research methodologies.* London: Sage.

CIMI. (2018). *Violence against indigenous peoples in Brazil 2017 data.* Report. Conselho Indigenista Missionário. https://cimi.org.br/wp-content/uploads/2018/12/Report-Violence-against-the-Indigenous-Peoples-in-Brazil_2017-Cimi.pdf. Accessed 25 May 2019.

Clifford, J. (2013). *Returns becoming indigenous in the twenty-first century.* London: Harvard University Press.

Couldry, N., Rodriguez, C., Bolin, G., Cohen, J., Goggin, G., Kraidy, M., Iwabuchi, K., Lee, K., Qiu, J. L., Volkmer, I., Wasserman, H., Zhao, Y.,

Koltsova, O., Rakhmani, I., Rincón, O., Magallanes-Blanco, C., & Thomas, P. (2018a). Media and communications. In *Rethinking society for the 21st century* (pp. 523–562). Cambridge: Cambridge University Press. https://www.ipsp.org/download/chapter-13. Accessed 25 May 2019.

Couldry, N., Rodriguez, C., Bolin, G., Cohen, J., Volkmer, I., Goggin, G., Kraidy, M., Iwabuchi, K., Qiu, J. L., Wasserman, H., & Zhao, Y. (2018b). Media, communication and the struggle for social progress. *Global Media and Communication, 14*(2), 173–191.

Coulthard, G. S. (2014). *Red skin, white masks.* Minneapolis: University of Minessota Press.

Custódio, L. (2018). *Favela media activism: Counterpublics for human rights in Brazil.* London: Lexington Books.

De Alcântara, M. L. B. (2019). Brazil. In D. N. Berger (Ed.), *The Indigenous World 2019* (pp. 134–142). Yearbook. The International Work Group for Indigenous Affairs. https://www.iwgia.org/images/documents/indigenous-world/IndigenousWorld2019_UK.pdf. Accessed 25 May 2019.

Deem, A. (2019). Mediated intersections of environmental and decolonial politics in the North Dakota Access Pipeline movement. *Theory, Culture and Society, 36*(5), 113–131. https://doi.org/10.1177/0263276418807002.

Dos Santos, G. L. (2006). *O Índio Brasileiro: o que você precisa saber sobre os povos indígenas no Brasil de hoje.* Rio de Janeiro: Laced Museu Nacional.

Downing, J. (2001). *Radical media: Rebellious communication and social movements.* London: Sage.

Downing, J., & Husband, C. (2005). *Representing 'Race'.* London: Sage.

Drugge, A.-L. (Ed.). (2016). *Ethics in indigenous research: Past experiences – Future challenges.* Umeå: Vaartoe – Centre for Sami Research, Umeå University.

Egan Sjölander, A. (2011). Introduction. In A. Egan Sjölander & J. Gunnarsson Payne (Eds.), *Tracking discourses. Politics, identity and social change* (pp. 13–48). Lund: Nordic Academic Press.

Emboava Lopes, C. (2016). *Silenced Realities: Exclusion of Guarani and Kaiowa perspectives in majority press discourse in Mato Grosso do Sul, Brazil.* MA thesis, Sámi University of Applied Sciences, Guovdageaidnu/Kautokeino.

Fairclough, N. (1992). *Discourse and social change.* London: Polity Press.

Fairclough, N. (1995). *Media discourse.* London: Arnold.

Fairclough, N. (2003). *Analysing discourse : Textual analysis for social research.* New York: Routledge.

Fairclough, N. (2017). CDA as dialectical reasoning. In J. Flowerdew & J. E. Richardson (Eds.), *The Routledge handbook of critical discourse studies* (pp. 13–25). London: Routledge. https://doi.org/10.4324/9781315739342.

Fairclough, N., & Wodak, R. (1997). *Discourse analysis: A multidisciplinary introduction.* London: Sage.

Foucault, M. (1971/1993). *Diskursens ordning*. Stockholm: Brutus Östlings Bokförlag Symposion.

Foxwell-Norton, K., & Lester, L. (2017). Saving the Great Barrier Reef from disaster, media then and now. *Media, Culture & Society, 39*(4), 568–581.

Foxwell-Norton, K., Forde, S., & Meadows, M. (2013). Land, listening and voice: Investigating community and media representations of the Queensland struggle for land rights and equality. *Media International Australia, 149*(1), 150–161.

Global Witness. (2018). *At what cost? Irresponsible business and the murder of land and environmental defenders in 2017*. Report. Global Witness. https://www.globalwitness.org/documents/19595/Defenders_report_layout_AW4_update_disclaimer.pdf. Accessed 25 May 2019.

Grünberg, G., & Melià, B. (2008). *Guarani Retã 2008 Povos Guarani na Fronteira Argentina, Brasil e Paraguai*. https://pib.socioambiental.org/files/file/PIB_institucional/caderno_guarani_%20portugues.pdf. Accessed 30 Aug 2019.

Guerra, V. M. L., & Valiente, M. F. (2014). Entre o discurso do "Acampamento Terra Livre" e a (ex)(in)clusão social, o (per)curso identitário dos povos indígenas. *Domínios de Lingu@gem, 8*(3), 86–107.

Gutierrez, Y. (2019). *A camera lens, a weapon in indigenous Brazil*. Global Landscapes Forum, Bonn. https://news.globallandscapesforum.org/36552/a-camera-lens-a-weapon-in-indigenous-brazil. Accessed 1 July 2019.

Huffington Post. (2019). Posse de Bolsonaro: Leia a íntegra do discurso do novo presidente no Planalto. *HuffPost Brasil*, https://www.huffpostbrasil.com/2019/01/01/posse-de-bolsonaro-leia-a-integra-do-discurso-donovo-presidente-no-planalto_a_23631426/. Accessed 13 Sept 2019.

IBGE. (2019). *Instituto Brasileiro de Geografia e Estatistica*. https://www.ibge.gov.br/apps/populacao/projecao//index.html. Accessed 23 July 2019.

Jørgensen, M. W., & Philips, L. (1999). *Diskursanalyse som teori och metod*. Frederiksberg: Roskilde universitetsförlag/Samfundslitteratur.

Kaiser, A. J. (2019, May 9). 'Exterminator of the future': Brazil's Bolsonaro denounced for environmental assault. *The Guardian*. https://www.theguardian.com/world/2019/may/09/jair-bolsonaro-brazil-amazon-rainforest-environment. Accessed 13 Sept 2019.

Kalapalo, T., Yawalapiti, A., Kuikuro, A., Kamaiurá, K., Waurá, A., Nafukuá, T., Matipu, A., Aweti, A., Mehinako, M., Trumai, K., Ikpeng, M., Suya, K., Yudjá, S., Kaiabi, M., & Associação Terra Indígena Xingu. (2019). *Carta de repúdio contra representação indígena na delegação do governo brasileiro na ONU*. Open Letter. https://coiab.org.br/conteudo/1569373095744x652187062137454100. Accessed 30 Sept 2019.

Kymlicka, W. (1995). *Multicultural citizenship: A liberal theory of minority rights*. Oxford: Clarendon Press.

Lawrence, R., & Raitio, K. (2016). Academia and activism in Saami research: Negotiating the blurred spaces between. In A.-L. Drugge (Ed.), *Ethics in indigenous research* (pp. 117–136). Umeå: Vaartoe – Centre for Sámi Research, Umeå University.

Lester, L. (2016). Containing spectacle in the transnational public sphere. *Environmental Communication, 10*(6), 791–802.

Levy, H. (2018). *The internet, politics, and inequality in contemporary Brazil: Peripheral media*. Lanham: Lexington Books.

Lindgren, S. (2017). *Digital media and society*. London: Sage.

Lindgren, S., & Cocq, C. (2017). Turning the inside out: Social media and the broadcasting of indigenous discourse. *European Journal of Communication, 32*(2), 131–150.

Londoño, E., Andreoni, M., Fountain, H., & Schwartz, J. (2019, August 24). What we know about the thousands of fires in the Amazon, the Planet's 'Lungs'. *The New York Times*.

Luke, A. (2002). Beyond science and ideology critique: Developments in critical discourse analysis. *Annual Review of Applied Linguistics, 22*, 96–110.

Markham, A., & Buchanan, E. (2012). Ethical decision-making and internet research: Recommendations from the AoIR Ethics Working Committee (Version 2.0).

Moore, E. (2019). *Journalism, politics and the Dakota Access Pipeline Standing Rock and the framing of injustice*. New York: Routledge.

Mussi, V. P. L., & Sousa, N. M. (2013). Povos Indígenas no Brasil – Outra Visão da História e da Literatura. In A. H. A. Urquiza (Ed.), *Culturas e História dos Povos Indígenas em Mato Grosso do Sul*. Campo Grande: Editora UFMS.

Napolitano, M. (2018). *The Brazilian Military Regime, 1964–1985*. Online encyclopaedia. Oxford Research Encyclopedia of Latin American History. https://oxfordre.com/latinamericanhistory/view/10.1093/acrefore/9780199366439.001.0001/acrefore-9780199366439-e-413. Accessed 11 Sept 2019.

Newman, N., Fletcher, R., Kalogeropoulos, A., & Nielsen, R. K. (2019). *Reuters Institute Digital News Report 2019*. Report. Reuters Institute for the Study of Journalism. https://reutersinstitute.politics.ox.ac.uk/sites/default/files/2019-06/DNR_2019_FINAL_0.pdf. Accessed 17 Sept 2019.

Ochigame, R., & Holston, J. (2016). Filtering dissent. Social media and land struggles in Brazil. *New Left Review, 99*, 85–110.

Oliveira, J. E. D., & Pereira, L. M. (2009). *Ñande Ru Marangatu: laudo antropológico e histórico sobre uma terra Kaiowa na fronteira do Brasil com o Paraguai, município de Antônio João, Mato Grosso do Sul*. Dourados: Editora UFGD.

Philips, T. (2019a, September 23). Brazilian indigenous leaders denounce Bolsonaro before UN speech. *The Guardian*. https://www.theguardian.com/world/2019/sep/23/jair-bolsonaro-un-brazilian-indigenous-leaders. Accessed 27 Sept 2019.

Philips, T. (2019b, September 24). Jair Bolsonaro says 'deceitful' media hyping Amazon wildfires. *The Guardian*. https://www.theguardian.com/world/2019/sep/24/jair-bolsonaro-says-deceitful-media-hyping-amazon-wildfires. Accessed 27 Sept 2019.

Pimentel, T. (2018, October 29). 'Ninguém solta a mão de ninguém', desenho que viralizou no país é criação de tatuadora mineira. *G1 Minas*, https://g1.globo.com/mg/minas-gerais/eleicoes/2018/noticia/2018/10/29/ninguem-solta-a-mao-de-ninguem-desenho-que-viralizou-no-pais-e-criacao-de-mineira.ghtml. Accessed 13 Sept 2019.

Quijano, A. (2000). Coloniality of power: Eurocentrism and Latin America. *Nepantla: Views From South, 1*(3), 533–580.

Rádio Yandê. (2019a). *Homepage Rádio Yandê*. http://radioyande.com. Accessed 23 July 2019.

Rádio Yandê. (2019b). *Rádio Yandê official Facebook page*. https://www.facebook.com/radioyande. Accessed 23 July 2019.

Rádio Yandê. (2019c). *Rádio Yandê official Instagram page*. https://www.instagram.com/radioyande. Accessed 23 July 2019.

Rádio Yandê. (2019d). *Rádio Yandê official Soundcloud page*. https://soundcloud.com/radioyande. Accessed 23 July 2019.

Rádio Yandê. (2019e). *Rádio Yandê official Twitter page*. https://twitter.com/radioyande. Accessed 23 July 2019.

Rádio Yandê. (2019f). *Rádio Yandê official Youtube page*. https://www.youtube.com/channel/UC9c_PSOqE9RCv5zG2Xb-5tA. Accessed 23 July 2019.

Ribeiro, D. (1995). *O Povo Brasileiro a formação e o sentido do Brasil*. São Paulo: Companhia das Letras.

Rodríguez, C., & El Gazi, J. (2007). The poetics of indigenous radio in Colombia. *Media, Culture & Society, 29*(3), 449–468.

Roosvall, A., & Tegelberg, M. (2015). Media and the geographies of climate justice: Indigenous peoples, nature and the geopolitics of climate change. *tripleC: Communication, Capitalism, Critique, 13*(1), 39–54.

Roosvall, A., & Tegelberg, M. (2018). *Media and transnational climate justice: Indigenous activism and climate politics*. New York: Peter Lang.

Salovaara, I. (2016). Participatory maps: Digital cartographies and the new ecology of journalism. *Digital Journalism, 4*(7), 827–837.

Sjölander, A. (2004). *Kärnproblem. Opinionsbildning i kärnavfallsdiskursen i Malå* (Core Issues. Opinion Formation in the Nuclear Waste Discourse in Malå). Dissertation, Umeå Studies in Media and Communication no 7, Umeå University, Umeå.

United Nations. (2007). *United Nations declaration on the rights of indigenous people*. Declaration. United Nations. https://www.un.org/development/desa/indigenouspeoples/wp-content/uploads/sites/19/2018/11/UNDRIP_E_web.pdf. Accessed 25 May 2019.

United Nations. (2016). *Report of the Special Rapporteur on the rights of indigenous peoples on her mission to Brazil.* Human Rights Council, United Nations. http://unsr.vtaulicorpuz.org/site/images/docs/country/2016-brazil-a-hrc-33-42-add-1-en.pdf. Accessed 25 May 2019.

United Nations. (2018). *Who are indigenous peoples.* Factsheet. United Nations Permanent Forum on Indigenous Issues. http://www.un.org/esa/socdev/unpfii/documents/5session_factsheet1.pdf. Accessed 10 Sept 2019.

Van Dijk, T. A. (2001). Critical discourse analysis. In D. Schiffrin, D. Tannen, & H. E. Hamilton (Eds.), *The handbook of discourse analysis* (pp. 352–371). Oxford: Blackwell.

Van Dijk, T. A. (2005). *Racism and discourse in Spain and Latin America.* Amsterdam: John Benjamins Publishing Company.

SECTION III

Digital Communication, Environmental Discourse and Local Experiences

Hacking the Foodscape: Digital Communication in the Co-design of Sustainable and Inclusive Food Environments

Alana Mann

Introduction

In 2013 George Packer suggested to readers of the *New Yorker* that government "is not a vending machine, with bureaucrats dispensing services, but a platform—like Facebook, Twitter and the iPhone—where citizens can build their own apps and interact with one another and come up with their own solutions" (Packer 2013). Indeed, there is widespread evidence that digital technologies are redefining the relationships and connections between citizens and governments. Employing social media, mobile computers, and data analytics global cities are engaging citizens in identifying, exploring, and contributing to the solving of social problems.

A. Mann (✉)
Department of Media and Communications, University of Sydney, Camperdown, NSW, Australia
e-mail: alana.mann@sydney.edu.au

J. Díaz-Pont et al. (eds.), *The Local and the Digital in Environmental Communication*, Global Transformations in Media and Communication Research - A Palgrave and IAMCR Series,
https://doi.org/10.1007/978-3-030-37330-6_9

Emerging technologies such as artificial intelligence (AI), virtual and augmented reality (VR/AR), robotics, digital traces, blockchain, and 3D printing are increasingly reshaping human action and have specific applications in involving citizens in co-designing healthier, more inclusive, and ecologically sound foodscapes. They are assisting cities around the world to tackle the "bads" of the food system through the reduction of food waste, the promotion of healthy eating, the consumption of products that are environmentally and ethically sound, and even the growing of local food. The creation of more innovative platforms and ecosystems that include citizens as explorers, ideators, designers, and diffusers is facilitated by tools that provide easy access to data, increased transparency, and reduce the distance between the citizen and the government agency (Nambisan and Nambisan 2013). Given the growing commodification of food for profit at the expense of our health and that of the planet (Mann 2019), closing this gap is essential to the creation of more democratic foodscapes.

Digital food activism (Schneider et al. 2018) is represented in a growing body of literature, but fewer connections have been made with the concept of digital citizenship (Witterhold 2018). This chapter seeks to address this issue by exploring how digital technologies can facilitate participatory engagement of ordinary eaters in urban food planning that connects cultural, socio-economic, and environmental issues and supports resilient and inclusive local food economies that counter the hegemony of Big Food. Describing inclusive and participatory local food initiatives in global cities, I present a perspective on the potential for citizens to identify their concerns in their local communities, construct their own agendas, and co-create healthier and more equitable food environments. This requires a cultural understanding of the role our food procurement and consumption habits play in citizenship. Considering the eater as a citizen rather than a consumer acknowledges the role of food as a "mediator of relations within social networks" (Grasseni 2017). This standpoint provides a basis for interrogating the capacity of digital technologies to join and scale across work in food systems transformation, which means addressing the structural inequalities that impact on individuals and communities experiencing "food violence" (Eakin et al. 2010) and tackling environmentally unsustainable food practices.

Food in a Changing Climate

Food is where the social, the technical, the cultural, the economic—and the environmental—meet. In 2008 food riots in over 40 countries were triggered by high food prices, lack of available food, and reactions against government food policies. Rising world market prices for food commodities, especially wheat, rice, soya, and maize, escalated the cost of staples for householders. Signalling a "critical stage in global neo-liberalisation" (Bohstedt 2014, p. 16) the riots were driven by citizens joining in popular movements against perceived and actual breaches of the social contract by political leaders. For those with a long view of global food and agriculture they were symptoms of a food system facing challenges that "are getting worse, not better" (The Food Ethics Council 2017, p. 6). These challenges emerge from a corporate-owned, industrialised food complex, colloquially known as "Big Food". Big Food relies on inequitable, high subsidies for the production of commodity crops in wealthy countries for profit. Overproduction provides surpluses dumped as food aid in poorer countries, exacerbating dependencies on food imports that have developed over the past 30 years with the failure of governments to invest in rural development and smallholder farmers (Golay 2010). Externalised costs in this system include loss of biodiversity, exhausted soils, declining water supplies, pollution, and extreme poverty.

The contradictions of modern agriculture are thrown into particularly "sharp relief" in the context of a warming climate (Taylor 2015, p. 193). The unequally distributed effects of climate change will be felt most severely in the areas where the hungriest and most malnourished live, including small farmers and rural communities where the majority of the world's food is produced (Holt-Giménez 2017; Mann 2014). Food systems contribute nearly one-third of our greenhouse gases (Gilbert 2012), while agricultural production including indirect impacts of land-use change contributes up to 80–86 per cent of total food system emissions (Vermeulen et al. 2012). The search for biomass alternatives to replace fossil-fuels, including corn and sugarcane for ethanol, drives some of these land-use changes which in turn take more land out of agricultural production. This triggers not only further stress on biodiversity but impacts heavily on the rural poor, who are often forced to migrate to cities. Urbanisation has grown from 10 per cent to 50 per cent of the world's population between 1990 and 2010 (United Nations Department of Economic and Social Affairs, UNDESA 2010). Urban areas, qualified as containing a minimum of 50,000 residents, take up just 3 per cent of the earth's surface but are responsible for approximately 70 per cent of global energy-related carbon emissions (Revi et al. 2014). While they are centres of intensive resource consumption, cities are also home to the food insecure even in affluent countries such as Australia (Dixon et al. 2011; Mann 2016, 2018b). With more than 50 per cent of the global population urbanised, and projected to increase to 66 per cent by 2050, it is critical that cities incorporate foodways into their planning for development and expansion.

Environmental and socio-economic risks, including extreme weather events, which have exposed the fragility of local and regional food supplies, have given rise to the concept of the "resilient city" (100 Resilient Cities n.d.). This discourse recognises that city and hinterland are interdependent and multi-scalar (McFarlane et al. 2016), comprising social, ecological, and technical systems that extend beyond city boundaries into, for example, sites of peri-urban agriculture. Resilience is an integrative construct that helps explain how people and communities achieve health and well-being. It is defined variously as "the ability to adjust in the face of changing conditions" (Pickett et al. 2004, p. 373); "the capacity to rebound from destruction" (Campanella 2006, p. 141); and "the degree to which cities tolerate alteration before reorganising around a new set of structures and processes" (Alberti et al. 2003, p. 1170). Driven by this necessity is the understanding of cities as sites for knowledge-creation or labs for innovation where "learning regions" and "clusters" ferment creative economies (McFarlane et al. 2016; McFarlane 2011). This positions citizens as active agents in devising creative solutions but tends to focus on economic innovation, regional competitiveness, and organisational learning rather than critically engaging residents in planning or addressing their dissatisfaction with power inequalities, absence of voice, and sense of exclusion (McFarlane 2011). How can we redefine the citizen's role in food systems planning by using digital infrastructure to engineer a shift from "passive service beneficiary" to "active informed partner, or co-creator" (Nambisan and Nambisan 2013) in food-related problem-solving?

Food Goes Digital

As new sociotechnical systems of distributed food production and consumption emerge, we are witnessing an "informational turn" in food politics (Frohlich 2017). This includes the growing reliance of consumers on digital media for information about food (Lupton 2018; Rousseau 2012) and the advent of digital food activism (Schneider et al. 2018). The successful digitisation of political consumption accelerates the adoption and diffusion of sustainable local food practices by individuals, while digital affordances that translate knowledge and make distant actors proximate facilitate transnational advocacy on food-related issues by creating political spaces for solidarity and resistance. In these spaces, activists connect, share strategies, and engage in networked framing (Papacharissi 2015, p. 75), build collective identity (Cottle and Lester 2011; Mann 2018a; Schneider et al. 2017), and engage in connective action (Bennett and Segerberg 2012; Benkler 2006). Linking their worldviews to local praxis, these actors collectively "identify a problem, propose a solution and motivate action" (Fox and Frye 2010, p. 425). Consumer netizens blur their online campaigning across issues such as genetic modification, food sovereignty, and Slow Food with political consumerism, lifestyle politics, and produsage (Bruns 2008). In many of these practices "giving voice to one's practice online is very important, not only in addressing others via a personal social web page, but also in searching for contact with like-minded people all over the world" (Witterhold 2018, p. 97).

The "flipside of virtual engagement" and its new forms of sociality is the way in which connectivity itself has become a "key source of value" for corporates exploiting audience labour and monetising data (Lewis 2018). Alongside the rise of digital food activism, Big Food has embraced Big Data and social media. Monsanto, now Bayer, has invested heavily in "be part of the conversation" campaigns and co-opted sustainable development through "legitimation strategies" (van Dijk 1998) that justify transgenic technologies as critical for the future-proofing of humanity against the multiple threats of climate change, food scarcity, and overpopulation (Mann 2019). Performing "open dialogue" with consumers through Twitter, Monsanto exploits the "myth of collectivity" (Couldry, cited in Lewis 2018, p. 193). In the processing and marketing sectors, global food corporations are "at the forefront of interactive marketing ... digital targeting and tracking techniques" and retailers have "taken over social media marketing companies" to access data and refine their marketing strategies (Stevens et al. 2016, p. 103).

The "digitisation of food" extends to automated, delocalised, and informatised processes of production and commercialisation of food including agricultural inputs where bioinformatics transforms seed genetic material into digital data, while robotics, drones, and other remote control technologies are revolutionising rural and urban food production (Filardi and Prato 2018). These "smart" farming technologies create efficiencies and increase profit margins under the guise of increasing production for enhanced food security but "often at the cost of environmental scrutiny and sustainable and equitable approaches to land management" (Lewis 2018, p. 195). On a city scale digital engagements in food systems often focus on procurement and distribution on web-based collaboration platforms to match supply and demand between small food processors and retailers and minimise the environmental impact of food production and distribution. Innovations include Milan's H2020 "U-Turn" project which connects farmers transporting produce into the city, and decision support tools to help stakeholders optimise regional food supply, as in the case of Rotterdam's FOODMETRES project (De Cunto et al. 2017). In California urban food system planning focuses on shifting everyday food practices and culture as well as developing technological innovation around data collection through tools like the Food Security in San Francisco: Dashboard created in 2015 to provide a centralised, comprehensive source of data on food-related programmes and activity and baselines on vulnerable populations such as the elderly, children, and low socio-economic groups. Urban agriculture (UA) is increasingly propelled by "ubiquitous technology, urban informatics and social media" (Hearn et al. 2014, p. 203) that facilitate the distribution and acquisition of food. As well as playing an increasingly important role in supplying food to cities, UA is providing stronger connections to and understanding of the food environment, and increasing awareness of the need to preserve small-scale and sustainable farming practices. For urban growers food-sharing applications like Ripe Near Me (https://www.ripenear.me/) facilitate on- and offline engagement and mitigate food waste, an objective shared by, OLIO (https://olioex.com/) and redirection apps like FoodForAll, Food Rescue Hero, NoFoodWasted, and No Food Waste, operating in countries ranging from the Netherlands to India. In the commercial kitchen platforms for the management and reduction of food waste such as Winnow, which uses Smart Meter Technology to log different types of waste produced in cooking, provide data that can lead to improvements in handling and procurement practices (Bozhinova 2018). In the consumption domain proliferating online review systems such as Urbanspoon and ingredient substitute search engines such as Fooducate assist individuals to make personal, and sometimes healthy, food choices (Urban Food Futures: ICTs and Opportunities 2014). Digitally supported forms of eco-labelling like QR codes and apps such a Buycott help us align our purchasing behaviour with our environmental ethics by providing information about point of origin, species, production, and consumption. All these "tactics of well-being" (Choi 2014) are double-edged—they provide us with information about our food in more transparent and systematic ways while revealing our food preferences to marketers.

Digitising Political Consumption

Digital networks are sources of counterpower in that they "enable communities to join for a common cause in opposition against industrial food production", providing "horizontal links" that "bypass" conventional channels and support the "social connections" that are vital to the emergence and sustainability of Alternative Food Networks (AFNs) (Stevens et al. 2016, p. 103). AFNs propose to reconnect with and reconstitute local human, cultural, and land ecologies as means to create and connect new spaces and models for engaging publics in debates over environmental sustainability, social justice, and economic viability, promoting a form of "food citizenship" and changing market relations (Gottlieb and Fisher 1996; Renting et al. 2012). Traditional forms of food activism, AFNs are now supported by websites, blogs, and social media which grow their consumer bases. Australian examples include the online organic delivery service CERES Fair Food in Melbourne (https://www.ceresfairfood.org.au/about-fair-food/); Food Connect in Brisbane (a social enterprise founded in 2005 by an ex-dairy farmer who aims to "democratise the food system") (https://www.foodconnect.com.au/about-us); and the rapidly expanding Open Food Network (https://openfoodnetwork.org/), a non-profit software platform that "connects and empowers food producers and hubs in regions or countries" and promotes "transaction and transformation". Comparable international initiatives include The Farmer: The Organic Store (https://thefarmeronline.com//home) based at Thakkar Farm on the River Page, 90 km from Mumbai, India; Farmer Uncle (https://farmeruncle.com/) in New Delhi; RadishBoya (https://

www.radishbo-ya.co.jp/shop/) in Japan; Tallo Verde (https://www.talloverde.com/) in Buenos Aires; and Club Organico (https://www.clube-organico.com/) in Rio de Janeiro.

Visiting these sites it becomes clear that accessing and purchasing healthy, organic, and/or sustainably produced food is more convenient and affordable for the digitally enabled, reproducing—and possibly augmenting—the disparities that exist in AFNs in the offline environment (Alkon et al. 2019). Digital technologies are potentially "exclusionary" either through cost, access to reliable network coverage, or the "outpacing" of "inclusive universal design features" by new innovations (Caldwell 2018, p. 25). They can "reproduce neo-liberal forms, spaces of governance, and mentalities" (Guthman 2008, p. 1171; see also Mann 2013), closing down possibilities for systemic and structural change by shifting responsibility from states and corporate actors to individual citizens (Foley 2018; Giraud 2018). Political consumption can generate "political quietism" (Lewis 2018, p. 197) in that it makes collective and direct action seem less necessary. Further, digital spaces are far from neutral, particularly when they become "sites for capital to extract value" (Lewis 2018, p. 196) whether it be through the exploitation of free labour produced by audiences or the monetisation of the data trails left by our online engagements. Yet "defining political consumption only as an economic, market-centric activity would mean disregarding much of the labour and creativity in which political consumers engage" (Witterhold 2018, p. 102) and ignoring the "new kind of solidarity" (p. 103) that can influence the consumption practices of eaters in positive, environmentally sustainable ways.

Local Food Governance

In food governance, as in other forms of soft governance, the emergence of "voluntary connections" through "orchestration platforms … that aggregate and evaluate the effects of polycentric initiatives while also creating opportunities for coordination, social learning, convergence, and reflexive adjustments towards continuous improvement" (Echersley 2019, p. 16) is evident. This rarely scales up beyond the local. The focus on city municipal governments is a rescaling of food governance that resembles a "retreat to localism" (Sonnino 2017, p. 1) but can actually drive transformative change. Good governance enables a government to deliver services to citizens efficiently through a combination of transparent and accountable institutions, strong skills and competence, and a fundamental willingness to do the right thing. Inclusive and participatory governance takes this a step further by drawing on the energy, influence, and knowledge of citizens in addressing common problems such as food insecurity. A more "flexible and porous approach to the re-localisation of the food system" (Sonnino 2017, p. 1), participatory food governance engages a large, active network of collaborative residents and professionals working together. This makes for a more resilient community than one led by heroic crusaders or reliant on short-term projects with restrictive budgets. In resilient communities residents suffering severe and multiple disadvantage, many of whom experience food insecurity, are more likely to recover and sustain their well-being through supportive social networks, as "wellness is created and lived by individuals every day through their ability to care for oneself and others, being able to take decisions, and having a sense of control … [factors that] depend upon strong social networks and high levels of social connectedness" (Civic Systems Lab 2017).

Building a robust participatory ecology requires public infrastructure including spaces for activities accessible to everyone, and enabling institutions such as food policy councils (FPCs). Ghent, Belgium, was one of the first European cities to have an FPC in which the city convened stakeholders to engage in joint efforts with civil society organisations and the private sector. These stakeholders, with collaborators across city departments, set multiple targets for food system interventions include climate change mitigation and adaptation, health, poverty, employment, health, and food waste (FAO 2018). In the US the number of FPCs has risen from 15 in 1999 to 324 in 2016 (Pothukuchi and Kaufman 1999; Sussman and Bassarab 2016). In some cases FPCs have played a significant role in brokering policy as a collaborative mechanism that pressures city governments to develop urban agriculture-friendly policies and programming (Morgan 2015). Notably, the influence of any FPC remains limited if left outside of the formal city governance structure and without stable funding. Some city planning authorities incorporate food into existing initiatives, such as healthy, sustainable, and resilient city programmes, while cities such as San Francisco in the US, Belo Horizonte in Brazil, and Milan, Italy, have dedicated municipal units overseeing urban food which demonstrates a high commitment to food as an essential infrastructure. Rotterdam Food Cluster, for example, comprises companies, education institutions, local governments, and residents engaged in a food sector that is the world's second largest exporter of agricultural products, employs 43,000 people in 8000 companies, and has an annual turnover of 25 billion Euros. The Food Cluster provides a platform for knowledge exchange, collaboration, investments, and innovation for food entrepreneurs across the entire food chain, including a Food Innovation Lab offering training for entrepreneurs to develop new business models of food production, storage, and distribution through the application of green technologies and the Internet of Things (Rotterdam Food Cluster 2017).

Novel approaches to harnessing the energy and creativity of citizens are now everyday practice for market leaders and governments. Examples include IBM's Innovation Jams to explore potential applications for new technologies, Fiat's crowdsourcing for futuristic car design, and Lufthansa's online competitions and games to gain insight into the customer experience. In the public sector the US is crowdsourcing solutions to "problems big and small" at Challenge.gov; the Danes are co-creating climate strategy through their Climate Consortium Denmark; and Finnish local governments are collaborating with senior citizens to redesign aged care. Mobile applications enable citizens to report local problems such as Boston's Citizen Connect, which closes the loop by reporting back to citizens on repairs to infrastructure. Beijing's open-source online database enables citizens to not only contribute in identifying factories violating China's green laws but facilitates their discovery of patterns in pollution. In Minnesota and Wisconsin in the US a similar approach engages residents in monitoring water quality. Data gathered by individuals is sent to responsible state authority. E-petitions provide channels for promoting policy change, as in the South Korean e-People initiative (Nambisan and Nambisan 2013).

These successful initiatives demonstrate that digital technologies work best when they enable roles for citizens in all stages of an issue cycle, including problem identification, solution conceptualisation, solution development and implementation. Going a step further are those that embrace the ethos of the participatory commons, such as the "food hackathons" held by GrowBot Gardens, at the San Jose Museum of Science and Technology, which encourage diverse groups to envision the future of small-scale agriculture through robotics and artificial intelligence (AI). The anarchist proposition that "no one person owns ideas or knowledge or solutions" (Caldwell 2018, p. 37) is also fundamental to the Every One. Every Day. project (https://www.

weareeveryone.org/) in Barking and Dagenham. This rapidly densifying and gentrifying East London borough where austerity cuts presented a "simple choice" to the municipal authority—"do nothing and continue to cut services, or can find new ways of delivering them" (Barking and Dagenham Growth Commission 2015)—entered into a partnership with the Participatory City Foundation to develop a participatory ecosystem, "a collection of many and varied 'participatory culture' projects" coupled with a support platform including "a design team, functional spaces, project ideas, organising, insurance, promotions, digital and metrics". Recognising that "innovative citizen-led local projects were achieving inclusive participation", the participatory culture approach encourages "common denominator activities" such as fixing, trading, recycling, batch cooking, growing, and making sustainable fashion, "experiences that were co-producing something tangible as a group of equal peers" (Civic Systems Lab 2017, p. 20). Adopting an "assets-based approach" that begins with "drawing together existing opportunities" in the community, the ecosystem is supported by digital infrastructure including government websites for transparency; online petitions and campaigning apps; and platforms for crowdfunding and crowdsourcing ideas (Participatory City 2018, p. 50).

The low/no cost, low commitment, and inclusive activities generated by the citizens participating in the project can be labelled "disruptive" in that they have the capacity to "transform food and food experiences in ways that make them accessible to as many people as possible, especially people who might otherwise be excluded from the food system as it currently exists" (Caldwell 2018, p. 37). They promote "cooperative efforts of mutual support and inspiration" that are made visible through digital platforms which collect participation and direct outcome data to measure impact of project activity including environmental and social benefits such as tons of food grown locally; square metres of land under cultivation; zero carbon journeys; hours spent in physical activity; and overall sense of happiness and well-being (Every One. Every Day. 2018). These platforms are increasingly useful in providing the public metrics and other data required to measure and drive the success of the project. This describes a politics of engagement that privileges learning in a place where residents experience diverse attachments, anxieties, and sometimes ambivalence about where they live. Urban geographer Colin McFarlane (2011) suggests this type of social learning be appreciated as a "distributed process" of translation across spaces that depends on the "(re)construction" of systems across different domains, and that the city itself is a "pluralised, multiple learning machine" (McFarlane 2011, p. 363) where "urban learning forums" embrace the "attached and cluttered" nature of the individual citizen and facilitate the reaching of "provisional understanding" (p. 370). These "hybrid forums" are sites where "questions and problems taken up are addressed at different levels in a variety of domains" (Callon et al. 2001, p. 18) by a diverse range of experts and lay persons.

Digital platforms facilitating this type of engagement offer possibilities for "progressive forms of learning between different constituencies" (McFarlane 2011, p. 360) if barriers such as access, education, and affordability are overcome. They might facilitate and sustain the emergence of "networked publics" (boyd 2010, p. 39). These "imagined collective[s] that emerge as a result of the intersection of people, technology and practice" enable people to gather and connect for social, cultural, and political purposes, interacting in unique ways that are determined by networked technologies that reorganise information flows, creating "emergent socio-political assemblages with shared or interlocking concerns" (Burgess and Matamoros-Fernàndez 2016, p. 81). They can provide marginalised groups with new tools to engage in tactical learning "to cope with, negotiate and resist" (McFarlane 2011, p. 360) the negative elements of the global food system.

Successful digitally supported initiatives involve partnerships between social actors including learning institutions, social enterprises, local government authorities, and residents who collaborate on doing everyday things, differently. According to Ezio Manzini (2011) a multiplicity of actors of this nature can set in motion initiatives that, from different points of origin, are moving collectively towards common ideals of well-being and sustainable production. "Small, local, open, and connected" in nature, these projects are establishing new relationships not only between production and consumption but also between the urban and rural nodes through, what is effectively, a "network of locals" (Manzini 2011, p. 102). Scaling *across*, rather than *up*, is the principle of translocalism in action.

Local, but Connected

Localism, according to Sonnino (2017), is inherently translocal. Expanding conceptions of local foodscapes beyond municipal boundaries has "important cross-scale repercussions" (p. 5), including learning and reflexivity in other cities. Convening around good practices and common conceptions of environmentally sustainable and equitable food production and consumption on a local level develops the capacity of all actors to move collectively towards greater system change. Translocalism has been adopted as a methodology for "scaling-out and scaling-up" creative solutions for more sustainable food systems featuring "citizens' involvement and social innovation, new forms of governance across levels of government, public procurement, and collaboration with research" (De Cunto et al. 2017, p. 128; see also City of Sydney 2017; Craven et al. 2018). Drivers for citizen-inclusion in urban food strategies include enhancing food security and nutrition; improving livelihoods of urban and peri-urban food producers; employment opportunities and economic development; the protection and restoration of local ecosystems; the reduction of the impacts of a changing climate; and increasing green spaces (De Cunto et al. 2017).

The Milan Urban Food Policy Pact (2015) is a good example of a "translocal urban learning forum" that translates ideas and models (McFarlane 2011). The 200-plus cities who have joined the Pact commit to engaging in peer-to-peer learning, dissemination of best practice and knowledge exchange in developing strategies, charters, and partnerships shared through digital platforms and at international meetings. In this context translocalism is a site for doing, performing, experimenting, practicing, and sharing things; for consolidating "networked relationalities" between food production and consumption (Sonnino 2017, p. 5). The Pact, emphasising "community buy-in and system thinking, and the preference for a multi-stakeholder approach", facilitates knowledge exchange within "a local and translocal network of cities around the world" (De Cunto et al. 2017, p. 89). Reducing food waste, promoting healthy diets, and respecting the environment, human rights, and worker's dignity were among the 100 concrete proposals for "local actions that can trigger global change" (Milan Urban Food Policy Pact 2015; see also FAO 2018) sourced via an intensive social media campaign engaging 1000 diverse actors (De Cunto et al. 2017, p. 89). These goals are supported by urban food planners and technology companies following the smart city agenda "to collect local narratives and to make powerful and socially embedded discourses emerge out of them" (Sharing Lab 2018). In terms of food systems these discourses extend beyond the local to the equitable sharing of the benefits and risks of how food is grown and processed, transported, distributed, and consumed; the provision of healthy food for all; ecologically sustainable food systems; fair conditions for food workers; decent profit for producers; and education to create informed eaters. In relating food systems to wider sets of public goods, new spaces of solidarity open up, enabling democratic capacity-building, possibilities for more sustainable practices, and an enhanced, cross-scale solidarity.

This form of progressive translocalism characterises the policymaking approach of a number of municipal governments faced with a "policy vacuum left by national policies entrenched within a larger scale productivist paradigm" (Sonnino 2017, p. 3). In her survey of 19 documents on food governance innovations in 13 cities across the UK, the US, and Canada, Sonnino (2017) identifies four primary values: a view of food as a multifunctional public good; an emphasis on civil society participation and governance; a flexible and inclusive approach to re-localisation; and a focus on the translocal. Local food plans in these cities share cross-sectoral approaches to food policy that unite climate action plans, transport and health and economic goals, and many emphasise the importance of building a health focused food system. The most progressive plans also reflect a clear recognition of the impact food consumption has on the environment. Systemic and participatory approaches emphasise the importance of the local scale, but "the re-localisation of the food system is never seen as an end goal; rather it is means to an end … part and parcel of wider sustainability strategies" (Sonnino 2017, p. 4). The role of the digital in these processes is not clear cut; they require imagination in the design of sociotechnical food futures that embody "more sustainable and equitable engagements with the digital realm" (Lewis 2018, p. 196).

Conclusion

Food activism, political consumption, and digital citizenship are undisputedly connected, requiring today's "food citizen" to develop a critical digital media literacy alongside food literacy. Digital technologies can provide an architecture for participation of citizens in collaborative co-design, mobilising "diffused social resources" including the creativity, skills, knowledge, and entrepreneurship vital in the transition to sustainability in our food systems. But "nothing can happen without the direct and creative participation of the people involved" (Manzini 2011, p. 103). Given that digital exclusion is, like food insecurity, an element of social exclusion we must take care "not to treat the Internet as the Holy Grail for environmental salvation" (cited in Lewis 2018, pp. 195–6). Food-centred practices and activities that occur in the physical communal environment can be "enhanced but not substituted by online communities" (Vallauri 2014, p. 173). Digital technologies serve to connect individuals across time and space on multiple scales and, arguably, some of their best applications lie closer to home. Inclusive and participatory approaches to the co-creation of healthy, sustainable, and vibrant foodscapes value food as a mediator in learning, as far more than just a commodity or simple nutrient sources. In this context "eaters are not just consumers but social actors whose meaning-making depends on faith, gender, age, income, or kinship" (Grasseni 2017). They play a vital role in peer-to-peer knowledge sharing and the adaptation of translocal ideas to local specificities. Food policy councils, urban learning forums focused on food, and small-scale citizen-led projects facilitated by digital mechanisms and infrastructure for co-creation including mobile applications, e-petitions, open-source databases, innovation jams, participatory design workshops, and dedicated online citizen communities have the potential to foster emerging forms of food citizenship and co-produce patterns of participation and belonging. We need more research into the "positive double link between grassroots users and technology" (Manzini 2011, p. 103) that might mediate the type of inclusive and participatory food governance that connects cultural, socio-economic, and environmental issues and sustains resilience in communities.

References

100 Resilient Cities. (n.d.). http://100resilientcities.org/challenges/food-insecurity/. Viewed 20 Apr 2018.

Alberti, M., Marzluff, J. M., Shulenberger, E., Bradley, G., Ryan, C., & Zumbrunnen, C. (2003). Integrating humans into ecology: Opportunities and challenges for studying urban ecosystems. *Bioscience, 53*(12), 1169–1179.

Alkon, A. H., Cadji, Y. J., & Moore, F. (2019). Subverting the new narrative: Food, gentrification and resistance in Oakland, California. *Agriculture and Human Values*. Available at https://link.springer.com/article/10.1007/s10460-019-09954-x. Viewed 20 June 2019.

Barking and Dagenham Growth Commission. (2015). *No-one left behind: In pursuit of growth for the benefit of everyone*. Available at https://ww2.lbbd.gov.uk/wp-content/uploads/2015/11/No-one-left-behind-in-pursuit-of-growth-for-the-benefit-of-everyone.pdf. Viewed 12 June 2018.

Benkler, Y. (2006). *The wealth of nations: How social production transforms markets and freedom*. New Haven: Yale University Press.

Bennett, W. L., & Segerberg, A. (2012). The logic of connective action. *Information, Communication & Society, 15*(5), 739–768.

Bohstedt, J. (2014). *Food riots and the politics of provisions in world history*. IDS working paper 444. Institute of Development Studies.

boyd. (2010). Social network sites as networked publics: Affordances, dynamics and implications. In Z. Papacharissi (Ed.), *Networked self: Identity, community, and culture on social network sites* (pp. 39–58). New York, Routledge.

Bozhinova, K. (2018). *Sixteen apps helping companies and consumers prevent food waste*. GreenBiz. Available at https://www.greenbiz.com/article/16-apps-helping-companies-and-consumers-prevent-food-waste

Bruns, A. (2008). *Blogs, Wikipedia, second life and beyond: From production to produsage*. New York: Peter Lang.

Burgess, J., & Matamoros-Fernàndez, A. (2016). Mapping sociocultural controversies across digital media platforms: One week of #gamergate on Twitter, YouTube, and Tumblr. *Communication Research and Practice, 2*(1), 79–96.

Caldwell, M. (2018). Hacking the food system: Re-making technologies of food justice. In T. Schneider, K. Eli, C. Dolan, & S. Ulijaszek (Eds.), *Digital food activism* (pp. 25–42). Abingdon: Routledge.

Callon, M., Lascoumes, P., & Barthe, Y. (2001). *Acting in an uncertain world. An essay on technical democracy*. Cambridge, MA: MIT Press.

Campanella, T. J. (2006). Urban resilience and the recovery of New Orleans. *Journal of the American Planning Association, 72*(2), 141–146.

Choi, J. H. (2014). Tactics of well-being: Mobile media and a new turn in the human-food relationship. In G. Goggin & L. Hjorth (Eds.), *Routledge companion to mobile media* (pp. 385–395). New York: Routledge.

City of Sydney. (2017). *Sustainable Sydney 2030: Community strategic plan 2017–2012*. Sydney: City of Sydney. Available at https://www.cityofsydney.nsw.gov.au/__data/assets/pdf_file/0011/288173/Adopted-Sustainable-Sydney-2030_Accessible-Version.pdf

Civic Systems Lab. (2017). *Designed to scale: Mass participation to build resilient neighbourhoods*. Available at https://drive.google.com/file/d/0B28SOnHQM5HVV0pyT2p1NGNvQk0/view. Viewed 10 Jan 2018.

Cottle, S., & Lester, L. (2011). *Transnational protest and the media*. New York: Peter Lang.

Craven, L., Scholsberg, D., & Mann, A. (2018). Addressing the SDGs through food business incubation: FoodLab Sydney. *Food & Cities: The role of cities for achieving the sustainable development goals* (pp. 145–151). Retrieved from the Barilla Centre for Food & Nutrition website: https://www.barillacfn.com/media/material/food_cities.pdf

De Cunto, A., Tegoni, C., Sonnino, R., Michel, C., & Lajili-Djalai, F. (2017). *Food in cities: Study on innovation for a sustainable and healthy production, delivery, and consumption of food in cities.* Brussels: Directorate General for Research and Innovation, European Commission.

Dixon, J., Beilin, R., Isaacs, B., Rose, N., Sanathanam-Martin, M., & Young, M. (2011). *Food security for the short or long term? Australia and New Zealand Agri-Food Research Network.* http://www.australianfoodsovereigntyalliance.org/blog/2011/06/06/food-security_short_long_term/

Eakin, H., Bohle, H. G., Izac, A. M., Reenberg, A., Gregory, P., & Periera, L. (2010). Food, violence, and human rights. In J. Ingram, P. Ericksen, & D. Liverman (Eds.), *Food Security and Global Environmental Change* (pp. 245–271). London: Earthscan.

Echersley, R. (2019). Ecological democracy and the rise and decline of liberal democracy: Looking back, looking forward. *Environmental Politics.* Available at https://www.tandfonline.com/doi/abs/10.1080/09644016.2019.1594536?journalCode=fenp20

Every One. Every Day. (2018). Available at https://drive.google.com/file/d/1K6xFXRLTXIedjmtI4vmv7a4eO7Mkc-7W/view. Viewed 20 June 2018.

FAO. (2018). *The role of cities in the transformation of food systems: Sharing lessons from the Milan pact cities.* Rome: FAO.

Filardi, M. E., & Prato, S. (2018). Reclaiming the future of food: Challenging the dematerialisation of food systems. *Right to Food and Nutrition Watch: When Food Becomes Immaterial: Confronting the Digital Age, 10*, 6–13.

Foley, R. A. (2018). Marketing critical consumption: Cultivating conscious consumers or nurturing an alternative food network on Facebook? In T. Schneider, K. Eli, C. Dolan, & S. Ulijaszek (Eds.), *Digital food activism* (pp. 110–129). Abingdon: Routledge.

Food Ethics Council. (2017). *Food citizenship: How thinking of ourselves differently can change the future of our food system.* Food Ethics Council. Available at https://www.foodethicscouncil.org/uploads/publications/FoodCitizenship2017_web.pdf. Viewed 11 May 2018.

Fox, R., & Frye, J. (2010). Tensions of praxis: A new taxonomy for social movements. *Environmental Communication, 4*(4), 422–440.

Frohlich, X. (2017). The informational turn in food politics: The US FDA's nutrition label as information infrastructure. *Social Studies of Science, 47*(2), 145–171.

Gilbert, N. (2012). One third of our greenhouse gases come from agriculture. *Nature.* Available at https://www.nature.com/news/one-third-of-our-greenhouse-gas-emissions-come-from-agriculture-1.11708

Giraud, E. (2018). Displacement, 'failure' and friction: Tactical interventions in the communication ecologies of anti-capitalist food activism. In T. Schneider, K. Eli, C. Dolan, & S. Ulijaszek (Eds.), *Digital food activism* (pp. 130–150). Abingdon: Routledge.

Golay, C. (2010). The food crisis and food security: Towards a New World food order? *International Development Policy* 1. Available at https://journals.openedition.org/poldev/145

Gottlieb, R., & Fisher, A. (1996). Community food security and environmental justice: Searching for a common discourse. *Agriculture and Human Values, 3*(3), 23–32.

Grasseni, C. (2017). *Food citizens: Collective food procurement in European cities: Solidarity and diversity, skills and scale*. Available at https://www.universiteitleiden.nl/en/research/research-projects/social-and-behavioural-sciences/food-citizens-collective-food-procurement-in-european-cities. Viewed 15 June 2018.

Guthman, J. (2008). Neo-liberalism and the making of food politics in California. *Geoforum, 39*(3), 1171–1183.

Hearn, G., Collie, N., Lyle, P., Choi, J., & Foth, M. (2014). Using communicative ecology theory to scope the emerging role of social media in the evolution of urban food systems. *Futures, 62*, 202–212.

Holt-Giménez, E. (2017). *A foodie's guide to capitalism: Understanding the political economy of what we eat*. New York: Monthly Review Press.

Lewis, T. (2018). Food politics in a digital era. In T. Scheider, K. Eli, C. Dolan, & S. Ulijaszek (Eds.), *Digital food activism*. Abingdon: Routledge.

Lupton, D. (2018). Cooking, eating, uploading: Digital food cultures. In K. LeBesco & P. Naccarato (Eds.), *The handbook of food and popular culture*. London: Bloomsbury.

Mann, A. (2013). Food sovereignty in a post-organic era. In B. Mascitelli & A. Lobo (Eds.), *Organics in the global food chain* (pp. 201–229). Ballarat: Connor Court Publishing.

Mann, A. (2014). *Global activism in food politics: Power shift*. Basingstoke: Palgrave Macmillan.

Mann, A. (2016). *The right to food and how 1.2 million Australians miss out*. Available at https://sydney.edu.au/news-opinion/news/2016/04/11/-the-right-to-food%2D%2D-and-how-1-2-million-australians-miss-out.html. Viewed 13 Apr 2018.

Mann, A. (2018a). Food sovereignty: Deep histories, digital activism and the emergence of a transnational public. In M. Phillipov & K. Kirkwood (Eds.), *Alternative food politic: From the margins to the mainstream*. Abingdon: Routledge.

Mann, A. (2018b). Hashtag activism and the right to food in Australia. In T. Schneider, K. Eli, C. Dolan, & S. Ulijaszek (Eds.), *Digital food activism* (pp. 168–184). Abingdon: Routledge.

Mann, A. (2019). *Voice and participation in global food politics*. Abingdon: Routledge.

Manzini, E. (2011). The new way of the future: Small, local, open and connected. *Social Space. Lien Centre for Social Innovation*, 100–105.

McFarlane, C. (2011). The city as a machine for learning. *Transactions of the Institute of British Geographers, 36*, 360–376.

McFarlane, S., Newell, J. P., & Stults, M. (2016). Defining urban resilience: A review. *Landscape and Urban Planning, 147*, 38–49.

Milan Food Policy Pact. (2015). Available at http://www.milanurbanfoodpolicypact.org/. Viewed 21 Feb 2018.

Morgan, K. (2015). Nourishing the city: The rise of the urban food question in the Global North. *Urban Studies, 52*(8), 1379–1394. https://doi.org/10.1177/0042098014534902.

Nambisan, S., & Nambisan, P. (2013). *Engaging citizens in co-creation in public services*. IBM Center for Business Development.

Packer, G. (2013, May 27). Change the world. *The New Yorker*, pp. 44–45.

Papacharissi, Z. (2015). *Affective publics: Sentiment, technology, and politics*. New York: Oxford University Press.

Participatory City Foundation. (2018). *Made to measure*. London: Participatory City Foundation. Available at http://www.participatorycity.org/made-to-measure-2

Pickett, S. T. A., Cadenasso, M. L., & Grove, J. M. (2004). Resilient cities: Meaning, models, and metaphor for integrating the ecological, socio-economic, and planning realms. *Landscape and Urban Planning, 69*(4), 369–384.

Pothukuchi, K., & Kaufman, J. L. (1999). Placing the food system on the urban agenda: The role of municipal institutions in food systems planning. *Agriculture and Human Values, 16*(2), 213–224.

Renting, H., Schermer, M., & Rossi, A. (2012). Building food democracy: Exploring civic food networks and newly emerging forms of food citizenship. *International Journal of Sociology of Agriculture and Food, 19*(3), 289–307.

Revi, A., Satterthwaite, D. E., Aragón-Durand, F., Corfee-Morlot, J., Kiunsi, R. B. R., Pelling, M., Roberts, D. C., & Solecki, W. (2014). Urban areas. In C. B. Field, V. R. Barros, D. J. Dokken, K. J. Mach, M. D. Mastrandrea, T. E. Bilir, M. Chatterjee, K. L. Ebi, Y. O. Estrada, R. C. Genova, B. Girma, E. S. Kissel, A. N. Levy, S. MacCracken, P. R. Mastrandrea, & L. L. White (Eds.), *Climate change 2014: Impacts, adaptation, and vulnerability. Part A: Global and sectoral aspects. Contribution of working group II to the fifth assessment report of the Intergovernmental Panel on Climate Change* (pp. 545–612). Cambridge/New York: Cambridge University Press.

Rotterdam Food Cluster. (2017). *Food Innovation Lab*. Available at https://www.rotterdamfoodcluster.com/2018/08/07/food-innovation-lab/?lang=en

Rousseau, S. (2012). *Food and social media: You are what you Tweet*. Plymouth: AltaMira Press.

Schneider, T., Eli, K., McLennan, A., Dolan, C., Lezaun, J., & Ulijaszek, S. (2017). Governance by campaign: The co-constitution of food issues, publics, and expertise thorough new information and communication technologies. *Information, Communication & Society*. https://doi.org/10.1080/1369118X.2017.1363264.

Schneider, T., Eli, K., Dolan, C., & Ulijaszek, S. (2018). *Digital food activism*. Abingdon: Routledge.

Sharing Lab. (2018). *Let's stop talking about smart cities*. Available at https://medium.com/we-research-and-expriment-with-how-the-sharing/lets-stop-talking-about-smart-cities-df5a0d187b51. Viewed 14 May 2018.

Sonnino, R. (2017). The cultural dynamics of food governance. *City, Culture and Society*. Available at https://www.sciencedirect.com/science/article/pii/S1877916617300139. Viewed 12 Mar 2018.

Stevens, T. M., Aarts, N., Termeer, C. J. A. M., & Dewulf, A. (2016). Social media as a new playing field for the governance of agro-food sustainability. *Current Opinion in Environmental Sustainability, 18*, 99–106.

Sussman, L., & Bassarab, K. (2016). *Food policy council report 2016*. Johns Hopkins Center for a Liveable Future.

Taylor, M. (2015). *The political ecology of climate change adaptation: Livelihoods, agrarian change and the conflicts of development*. Abingdon: Routledge.

UNDESA. (2010). *World urbanisation prospects: The 2009 revision highlights*. New York: UNDESA.

Urban Food Futures: ICTs and Opportunities. (2014). *Futures, 62*, 151–154.

Vallauri, U. (2014). Transition Belsize Veg Bag scheme: The role of ICTs in enabling new voices and community alliances around local food production and consumption. *Futures, 62*, 173–180.

van Dijk, T. A. (1998). *Ideology: A multidisciplinary approach*. London: Sage.

Vermeulen, S. J., Campbell, B. M., & Ingram, J. S. I. (2012). Climate change and food systems. *Annual Review of Environment and Resources, 37*, 195–222.

Witterhold, K. (2018). Political consumers as digital food activists? The role of food in the digitalisation of political consumption. In T. Schneider, K. Eli, C. Dolan, & S. Ulijaszek (Eds.), *Digital food activism* (pp. 89–109). Abingdon: Routledge.

Smart City and the Reinvented Politics of Governing Through Datafied Environment in Songdo, South Korea

Chamee Yang

This chapter examines the linkage between two governing strategies: the discourse of sustainable development and sociotechnical infrastructure of smart cities. Drawing from a case study of Songdo, South Korea, this chapter problematizes the productivity of the nexus between these two policy universes—by unpacking the interplay between the historical evolution of sustainable development as a new growth paradigm and the new urban governance model of smart cities afforded by the digital technologies, such as protocols, sensors, and urban data platforms. In doing so, this chapter directly engages with the volume's theme, "When the Local Meets the Digital," by interrogating the role of datafication in the local experience of a "smart and green city" in South Korea.

C. Yang (✉)
Institute of Communications Research, University of Illinois at Urbana-Champaign, Urbana, IL, USA
e-mail: cyang62@illinois.edu

J. Díaz-Pont et al. (eds.), *The Local and the Digital in Environmental Communication*, Global Transformations in Media and Communication Research - A Palgrave and IAMCR Series,
https://doi.org/10.1007/978-3-030-37330-6_10

This chapter begins by tracing a brief history of Songdo's reinvention as a "smart" and "green" city and investigates the social and historical conditions of current discursive formation. It observes how the city's strategy to reposition itself is articulated through its new urban design scheme and looks at the programmes that responded to the citizens' increasing demand for a higher quality of life. As will be shown, this transformation of Songdo did not only involve a linear passage of economic policy from an industrial to a post-industrial one. Instead, the city was brought into an emerging network of relationships among diverse actors and things including the nation-state, IT corporations, transnational non-governmental organizations, digital technologies, the urban environment, and the citizens.

This chapter also attends to how the transition was largely driven by the rationale that posited sustainability as an aim of a more ethical development than the previous ones and the consensus that "responsibilized" the citizens for protecting nature and future generations, as well as themselves. In this way, the hybrid relationship between ethics and the economy foregrounded the civic virtue demanded of the globally minded, ecologically conscious, and responsible "good citizen." More specifically, this chapter addresses how the "smart and green city" reflects the strategies of governing through ethics and the normative way in which the government operates in contemporary cities. The primary object of this investigation is the institutional arrangement and discursive landscape that conditioned the technical development and distribution of the smart city as a popular idea and a solution to the problem. This chapter approaches the analysis from the perspective of "problematization" (Koopman 2013): that is, there is an intricate process in which the sustainable development agenda established the environment as the problem, and the smart city proponents appointed themselves as offering one of the solutions to the problem.

Beginning with the case study of Songdo, this chapter challenges the role of the digital in the local experience of smart and green city, in the sense that the digital technologies were used to replace the local problem with the universal problem; that the digital served as a technical fix to the problem that itself defined; and that the digital became a crucial means of assessing, comparing, and governing contemporary cities.

Reinvention of Songdo, the Smart and Green City

Since its early conception in the late 1980s, the proponents of the Songdo development project on the Incheon urban development committee stressed their commitment to pursuing the vision of a global and high-tech city that would offer a higher quality of life. Park Yeon-Su, Incheon's urban planning chief in 1986, drew a plan for the "northeast Asian international business center city project," which later became the blueprint for Songdo's development. The plan consisted of "applying cutting-edge technology into real life to maintain the environment, transportation, and security so that one can communicate knowledge, information, and culture anytime, anywhere" (Park 2008, p. 100). It also involved designing an "urban environment that possesses ample green space, water, and architectural aesthetics" (ibid.). Ultimately, the plan envisioned a city that pursues "a multicultural society through deregulated and global business activities" (ibid.).

The ambitious vision of the government-initiated Incheon Public Development Center was a response to growing anxiety fuelled by the rapidly changing economic and political landscape of the East Asia region. The hope was that Songdo's development into an international business district and information hub would renew regional pride in Incheon. For a city that had been South Korea's major industrial production site, Songdo was envisioned as a city's new model for its future, which would bring in new business opportunities and sources of income. Besides, the growing manufacturing industry in China had been adding insecurity for the city that had mainly relied on the heavy and chemical industries, with a majority (77.2%) of its workforce involved in the manufacturing industry. The abundance of cheap labour in China, once opened up to the international markets, would make Incheon's and South Korea's relative advantage of skilled labour increasingly untenable (Park 2008, p. 26).

Since 2007, the publicity and marketing materials for Songdo started to shift in tone by promoting Songdo as a "sustainable," "green," and "eco-friendly" city (Kim 2010). This change of branding strategy again reflected both the changing national and international political and economic conditions. The global financial crisis in 2007 redirected real estate investors and financial institutions to a more cautious approach towards investments. Even though residential sales in Songdo were steadily growing, the developers hoped to increase their competitiveness by spotlighting its sustainable design. The national economic policy under President Lee

Myung-bak's (2008–2013) regime adopted a low-carbon, "green growth" strategy. City officials of Songdo and Incheon Metropolitan City adapted to these conditions and proclaimed their willingness to reshape their urban policy in conjunction with the national policy goal for green growth.

Songdo was an exemplary case for demonstrating the city's commitment to the value of sustainability through an array of technologies and the reconfiguration of the urban landscape and infrastructure. For instance, "G-tower," one of several landmark office buildings in Songdo, changed its name from "I-tower," which signified the city's transformation from international business hub to a technological zone more comprehensively encompassing the values of "green," "global," and "growth." In addition, Songdo allotted 40% of its land to green space circumscribed by cycle paths and walking routes. The advanced bike infrastructure in Songdo, as well as a higher rate of bike use than any other districts in Incheon, seemingly corroborated the fact that greenness and sustainability had already taken centre stage in daily life in Songdo. Real estate developers and city officials often promoted the walkable, bike-friendly, and green urban landscape of Songdo that evoked Ebenezer Howard's Garden City image, aiming to attract people distressed by the busy, congested, bustling life in the metropolitan city. Events like the annual Bicycle Festival inaugurated in 2012 were organized by the city that hoped to attract many tourists and visitors seeking a moment of calm and tranquillity outside of Seoul.

Technical arrangements of the smart city further enhanced the city's goal to "go green." Electric vehicle charge stations and smart grid, smart home, and automated infrastructure services consisted of digital technologies deployed to centrally manage water, electricity, transport, and garbage disposal in order to optimize the city's overall energy consumption and lower maintenance costs (Songdo IBD 2013). A smart grid system is a networked technology that allows two-way interaction capacity built into the power utility, designed to detect the peak hours of electricity use and react to individual demand for electricity. Thanks to this technology, Songdo was said to be emitting fewer greenhouse gases, and the residents were said to be using 40% less energy than those in average cities. Smart homes, equipped with digital control panels and multiple interior sensors, allowed people to control lighting, heating, ventilation, and electricity use at home, to view their monthly energy usage, and to compare it with everyone else's (James 2016).

Sustainability has also been figured as the fundamental architectural design principle in Songdo's iconic buildings. The most exemplary case is

Songdo Convensia, a convention centre that acquired certification from the Leadership in Energy and Environmental Design (LEED) programme, becoming the first in Asia in 2009. Stan Gale, the CEO of Gale International, the real estate development company that master-planned Songdo with officials in the Incheon Free Economic Zone, made a self-congratulatory remark on this recognition and confirmed his commitment to "foster environmental awareness" (Yoneda 2010). The signification of Songdo Convensia's acquisition of LEED certification was hoped to "engender a real sense of eco-responsibility within the larger Songdo IBD community" (ibid.). Officials of Songdo took pride in the fact that their city became a "showcase to the world as a major example of sustainable development and green growth" (Yi 2012).

LEED is a flagship programme of the US Green Building Council (USGBC), a Washington, DC-based non-profit organization that has promoted sustainability in building design and has created a buzz around green building since 1993. LEED consists of the list of parameters to rate a building's "environmental performance." Songdo utilized the LEED standard for several building designs and, as a result, has the highest number of LEED-certified buildings among cities in Asia. The LEED sustainability rating and its certification system is currently the most widely used standard across the US and the rest of the world; many municipalities and corporations have used LEED as a plausible standard for certifying and approving the greenness of their building or city. Such a broad appeal of standardization like LEED reflects how "going green" has become a powerful incentive and a prerequisite for construction companies driven by the potential benefits of having an ethical image.[1]

LEED's initial parameters for the building included the availability of bicycle facilities, storage and collection of recyclables, the level of indoor air quality, water metering, and energy performance. The US Green Building Council was also on its way to develop its locally-customizable LEED design standard for the cities around the world, which was published in December 2016. In 2017, Songdo was again awarded the certification for the LEED for the Cities programme, the first in Asia, joining ranks with other certified cities including Phoenix (AZ), Arlington (VA), and Washington, DC. LEED's standardization for cities was developed as

[1] In the US, more than 60,000 commercial projects were participating in LEED, and this system was said to have spawned an entire "green building" industry worth up to approximately $248 million in the US by 2016 (US Green Building Council official website).

a template for cities to set performance goals, track their performance through data, and measure progress towards their goals.[2]

Through willing participation in a standardization programme such as LEED, data have become the primary means to track the environmental performance of Songdo, the new smart and green city. They promote a computational way of viewing and assessing the city based on the universal criterion of the carbon footprint. Songdo is a prime example of cities that readily adopt the data-enabled comparative urbanism—the idea that the environmental performativity of a city can be assessed through data and that each city can compare its progress towards the universal goal with other cities in the world. The global trope of comparative and competitive urbanism is broadly spreading across many other cities in South Korea and beyond.

Yet it would be reductionist to assert that highly motivated real estate developers and the city government striving to put the city on the map were the only ones accountable for giving emergence to the smart and green city of Songdo. Strategies of the government to problematize the environment through the discourse of sustainable development responded to its recognition of a demand for a higher quality of life and a role that the clean environment might play for its citizens. It also involved the government's own reflection on the older practices of urban planning predicated on the industrial economy, which necessitated a significant update.

A Smart and Green City for a Better Quality of Life for Citizens

The environmentally-oriented urban policy in Songdo stated above mostly charted the technologies of urban infrastructures and programmes to manage them. In a broader scope, however, these technologies are intricately tied with the mandate for sustainability that grew out of the demand for a better quality of life in the city that has a clean and safe neighbourhood, as well as the government's evolving strategies to *govern through the environment*.

[2] According to Vatsal Bhatt (2016), the director of Cities and Neighborhood Development at USGBC, the new LEED for Cities programme asks cities to provide data across five categories to generate a performance score: energy, water, waste, transportation, and human experience. The human experience category includes various sociological data such as the percentage of the population with a high school diploma or bachelor's degree, median gross rent and household income, crime rate, unemployment rate, and so on.

Songdo's residential sales have grown steadily despite the financial crisis of 2007 that stymied several development projects (Kim 2010). One of the reasons behind Songdo's growth was the government's support for companies relocating to Songdo, which included providing tax incentives and selective deregulation from policies that were applied elsewhere in the country (Park 2005). Another reason was that Incheon and South Korean middle-class families responded favourably to Songdo's well-promoted features of security (e.g., CCTV and control tower, the lowest crime rate in the country); health (e.g., walkable neighbourhood, bike paths, public parks); education (e.g., English-speaking private international schools, global campuses launched by American universities); and leisure (e.g., iconic landscape, galleries, shopping malls).

The popularity of Songdo reflected the increasing prominence of the quality-of-life discourse. The quality-of-life discourse became especially prominent in the twenty-first-century South Korea, responding to the country's internal reflection on the mode of growth during the country's rapid industrial period and its supposed lack of concern for culture and the environment. Modern urban planning that began in the 1960s was often represented through images of smokestacks from industrial complexes and matchbox-like apartment complexes that suppressed individual dignity. The city had always been treated as a medium for economic growth, national pride, and technical achievement in the modern urban development of South Korea (Lee 2002). The newly perceived challenges of globalization recognized the city as *a problem space*, wherein the provision of a clean, pleasant, and upscale living environment was integral to sustaining growth and to its aspiration to become a metropolis presentable to the world market (Wilson 2007). Modern industrial cities, which had often been associated with the disorderliness of the crowd and outdated infrastructure that posed a threat to public health, were cast as the problem in need of reinvention through the emerging discourse of the smart and green city. This problematization helped establish the governmental aim to reshape and redesign the urban landscape as a "network of living practices of well-being" (Osborne and Rose 1999).

Songdo, as well as many other proliferating new towns in South Korea, readily adopted the new mode of governing that involved reshaping the relationship between the city and the environment, wherein the environment was redefined as a governmental problem in need of regulations, programmes of action, and policies. In this sense, Songdo's reinvention as a smart and green city is a response to the changing national and international political and economic conditions and a demand for a higher quality of life of citizens. Through encouraging micro-practices such as opting for paperless billing, replacing plastic bags with eco bags, recycling garbage, and "bringing your own cup," citizens' willingness and commitments towards green consciousness were mobilized for collectively making a good city. Morally charged and responsibilized citizens embraced the civic virtue of protecting the environment, which was understood as a way of improving their own quality of life. It is through this ethical elevation of the environment as an objective and an instrument of the government that the mandate for sustainability becomes more than just a government initiative, but a collective goal worthy of bringing grassroots communities, non-governmental organizations, and private businesses to the same line of actions.

It is important to note the sense that "we are all in this together" is predicated on conjuring the planetary scale of a threat that crosses regional boundaries. The individual citizen, as the basic unit of action accountable for practicing good citizenry conduct and protecting the environment, is imagined as an undifferentiated, universal subject. This reconceptualization of the individual as bearing the responsibility of global scale displaces the subject from the conventional political landscape organized by the hierarchy of the nation-state. Both sustainable development and the smart city, as will be discussed in the following section, are implicated in conjuring the global sense of time and space governed by technical standards and datafication, which replace the conventional language of politics.

Sustainable Development as the Universal Problem and the Technical Fix of the Smart City

As seen in the case of Songdo, the imperative for smart and green city development gained profound impetus from establishing the environment as the government's problem; mobilizing the citizens' responsibility and demand for quality of life; and evoking a global sense of time and space that transcended the boundary of the nation-state. To substantiate the above arguments and to more closely examine the relationship between "the digital and the local," it is useful to think through the role of technology in defining the problem it is brought in to solve.

To assert that digital technology is redefining the problem of the environment is another way of saying that technology is offered as a solution to the problem. In response to the environmental problem induced by massive industrial production, digital technology assumes the position of an alternative solution, notwithstanding the environmental damage that it has, itself, caused (see Maxwell and Miller 2012; Gabrys 2013). More importantly, however, not only is technology addressing the problem it claims to solve but it is also defining the problem in its own terms. In other words, technology has been implicated in both creating the problem and solving the problem.

However, the technologies of the smart city, like any other technologies, can only be made operational by responding to existing social demands and solving problems. Then, it is crucial to inquire how the problem was problematized. As seen in the case of South Korea, the vision of the smart and green city was formed as a response to the criticisms levelled against the modern industrial economy. The government's subscription to the emerging economy of data-driven governance signified a departure from the industrial model of development that relied on fossil fuels and oils and a turn to data as the new carbon-free growth factor. To fully investigate the political implication of this transition and the relationship between the smart city and sustainable development, it is essential to trace the historical evolution of each of these concepts.

Smart City as a Solution to Environmental Problems?

The smart city, as technological innovation and a vision for urban future, has gained traction among urban policymakers and IT-related business circles as of late, especially since around the 1990s (Albino et al. 2015). This futuristic terminology first appeared in a report published by the California Institute for Smart Communities at San Diego State University, entitled "San Diego: The City of the Future—the Role of Telecommunications." In 1994, as part of their commission for the City of the Future initiative, the Institute published a report calling for an urgent need to build "digital communities" along informational highways. The idea was to streamline every home, office, and school with wired and wireless connections. Evoking a compelling sense of the reality of the so-called knowledge economy, the report also emphasized the importance of collaboration between university research centres and the business community, and the need to reinvent the city to become smart. The proposed policy direction later became the basis for the Smart Communities programme implemented in the state of California in 1996 (Eger 2005).

This new model of the city, presented as a logical outcome of history, corresponds to an administrative desire to rationalize and control the city and is now presented as an alternative mode of development. Crucial to legitimatizing this vision of an advanced future is the role of technology as the key enabler. Many existing, in-progress, and soon-to-be-built smart cities have highlighted the investment in new information and communication technologies (ICTs) that would facilitate the efficient governing and development of the city. Although different cities have had different levels of motivation and expectation, what they all had in common seemed to be the expectation for technology to aid the government's aim—be it economic revitalization or efficient urban management (Vanolo 2014).

Since 2006, Cisco, a Silicon Valley-based corporation that produces network hardware and telecommunication equipment, has actively established itself as a business that incorporates the value of sustainability. Cisco came on board with the Songdo development project in 2009 and used the opportunity to materialize its vision for *Smart+Connected Communities*, which was to utilize the information network as an underlying platform for urban infrastructure, managing utilities, transportation, security, and governance. Primary benefits, it was said, were increased energy efficiency, economic opportunity, and quality-of-life gains for citizens (Cisco 2010). To date, Cisco's public relations strategy connects their new developments in smart city and Internet of Things (IoT) technology, which enables automated heat, gas, and energy monitoring, as well as remote collaboration that reduces business travel, with the environmental benefits of going sustainable (Firger 2017).

In Songdo, smart city technologies have been deployed in the context of environmental problems and other pressing problems in the areas of security, mobility, and economic growth. Songdo, as well as many other cities that are participating in the smart city development programme, emphasizes the new technical features that will help improve existing infrastructure and services and meet the goal of sustainable development. Such instrumental approach to the smart city is becoming prevalent in many other smart cities development where technology is seen as the crucial solution to the environmental problems. In this way, the proponents of the smart city who view the world as connected, networked, and programmable space find a new expression in the discourse of sustainable development, through which the environmental problems come to matter in different ways. The problem of environment in sustainable development, as it will be seen in the following section, becomes subjected to the alternative economic growth strategy and the governmental aim of datafying and ordering the whole of urban environment and population.

Sustainable Development, Accumulation, and Depoliticization

It was no small coincidence that the idea of the smart city was concurrently formulated with the increasing imperative of sustainable development in South Korea. Sustainable development was first proposed by the World Commission on Environment and Development (WCED) and its report, also known as the Brundtland Report entitled, "Our Common Future," published by the United Nations in 1987. The report called for the cooperation and commitment of all countries to conceive an alternative pathway of development to meet the basic needs of all; to extend opportunity; and to fulfil aspirations for a better quality of life. Sustainable development is herein defined as the "development that meets the needs of the present without compromising the ability of future generations to meet their own needs" (WCED 1987), culminating in the first Earth Summit in Rio de Janeiro in 1992. While every country was unilaterally enrolled in the challenge to offset the anticipated crisis of the future, the political will of developing countries was regarded as especially important to improve their inadequate infrastructure. To compensate for the lack of resources and expertise in developing countries, The World Bank and the International Monetary Fund (IMF) were appointed as the conduits to help finance the project and implement policies in those regions.

Although sustainable development was not only a problem of growth and capitalism, the concern for economy and production constituted a significant part of its imperative. Historical evolution of this concept reveals that sustainable development is largely a response to the problem of growth. The fear of resource obsolescence was instigated by the Club of Rome's publication of its landmark report on *The Limits of Growth* in 1972, which predicted that exponential economic and population growth would deplete the finite resources available on the earth. This alarming simulation was a wake-up call for some businesses and economists. Public paranoia, followed by the oil crisis in the late 1970s, reinforced the need to take seriously the possibility of limits to growth and prosperity.

A more careful examination of this concept reveals a tension between "the sustainable" (denoting stasis) and "development" (denoting movement), which caused much confusion and debate (Redclift 2005). First of all, one can imagine the tension between those who pointed to industrial production as the root cause of ecologically destructive practices and those who thought that capitalism could be made sustainable if the metabolism problem was overcome through technology. For the mass industrial production that depended on scarce resources (e.g., oil, gas, human labour), technology was a vital means to maximize the profit from such scarce resources. In the wake of the climate crisis and the fear of limits to growth, the choice was presented whether to keep polluting the environment or switch to non-scarce natural resources. It was in this context that sustainable development became the topic of an alternative accumulation strategy that seemingly reconciled the conflicting desires—between the human-centric preference to continue economic growth and the eco-centric perspective to value ecology as an end in itself (Haque 2006).

In South Korea, the proponents of sustainable development and green growth were less concerned with protecting the environment than with choosing the next model of wealth accumulation. "Data is the new oil," the often-propagated slogan for the emerging economy of sustainability, demonstrates this point. From clean energy industry to IoT network infrastructure production, the whole of smart technology industry relies on an overriding agenda of sustainability to justify the data-driven growth model.

That being said, to view sustainable development only as an alternative accumulation strategy is to risk a great reductionism. There exists a more serious mechanism of power at play here, which involves dissolution of politics into the matter of technical assessment and progress. While the anxiety over the limits to the capitalist accumulation regime lies behind the continuing challenges to improve and redesign less energy-intensive and more cost-effective urban infrastructure, sustainable development also affords a legitimation for imposing a regulatory mechanism that pervades the city and the environment. In this sense, the sustainability mandate is closely tied to the computational means of managing, monitoring, and controlling the city and the environment, which dodges political accountability to the algorithms.

Political Implications of the Wedding of the Smart City and Sustainable Development

At Greenbuild Expo in Los Angeles in 2016, many small and medium-sized startup companies from Silicon Valley showcased their new bundles of software and services to cater to the desires of many cities to become smart and sustainable. The event was one of many proliferating venues where the buzzwords of smart cities and the Internet of Things (IoT) are tirelessly reproduced and circulated among city officials, businesses, and policymakers. Although the initial investment may be costly, in the long run these technologies are said to reduce the city's spending on electricity and fuel consumption considerably.

In such a highly technocratic approach favoured by municipalities and corporations, the bottom line of their proposals always touch upon adopting new zoning rules, building codes, and protocols. Carbon footprint is the primary measure of tracking a city's progress. Organizations, such as the US Green Building Council (USGBC) and American National Standards Institute (ANSI) that are now developing the standardized framework for smart cities, work closely with governments and IT corporations (e.g., Cisco and Intel), to further consolidate the relationship between sustainable development and smart cities.

One of the consequences of wedding the politics of the environment with the technical means of smart cities is that the matter of politics is relegated to the matter of computation and statistics. Problems of social relations are quantified and reduced for the sake of making progress towards the common goal, notwithstanding the fact that many developing countries have different policy priorities, and some may not even be able to purchase new technologies to improve their infrastructures or defer the immediate reward of carbon-dependent economic growth for the sake of the long-term reward of a clean environment.

To repeat, the political implications of wedding smart city and sustainable development stated above are the displacement of sociopolitical actions by data science, which datafies the environment and justifies this through ethics and the global commons. Traditional political boundaries of the state and civil society are translated into the language of technology, ecology, and the whole of the planet. In other words, in the smart city where everything is seemingly datafied and programmable, complex social relations that pertain to the environmental problems are casually reduced and reified to a list of environmental performativity metrics, which then can be technically modified.

More importantly, this emerging paradigm of governance through datafication of the environment inducted cities into the zone of practicing ethical means of development. Here, the environment is treated both as an object of the government, which has to be protected, and as an instrument of the government, which has to be datafied and controlled. Cities and citizens are implicated in the universal application of these ethics for the anticipation of a better life, better future, and better self, although the actual environmental impact that the technical application of smart and green city technology accomplishes is little to non-existent—for instance, one can score LEED points without actually doing much good for the environment (Udall and Schendler 2005).

Lastly, the fundamental presumption of the smart city-enabled sustainable development vision is the notion of the "global commons." As the argument goes, everything is interrelated: the problems of developing countries are problems for developed countries; nature and the economy are inseparable; and it is vital to act as a global commons to prevent ecological and economic disaster. The invocation of environmental disaster anticipated by ongoing climate change is not only portrayed as a global crisis in which the whole of nature is implicated, but is constituted as a universal humanitarian threat, which supposedly produces people as the universal political subject and the whole of humankind (Swyngedouw 2010). Global commons supposedly transcend national borders—which is why a universal standard such as the LEED programme is gaining prominence as a guideline.

This new mode of governance, however, has limitations for at least two interrelated reasons. First, the wholeness of the global commons transcends the scale of ordinary citizens, cities, and countries, and yet privilege is given to the specialists who do not belong to the politically demarcated territories (Latour 1998). Ecology integrates itself into everyday life, yet it becomes "inflated to the point of assuming responsibility for the agendas of all" while giving advantages to the group of specialists who speak on behalf of "a global unity which no longer has the political domain as its horizon" (Latour 1998, p. 3). For instance, whose "needs," exactly, are being addressed in the sustainable development imperatives, and who exactly is going to benefit from the next development path, is uncertain (Redclift 2005, p. 214). It is casually assumed that the idea of sustainability is unquestioningly good and that everyone's stakes are involved in it; therefore, all countries (and cities) must contribute to tackling this imminent challenge on a global scale. Although there has been little consensus about how the needs of different countries could be defined differently and fewer concerns about the fact that not everyone could afford the privilege to "choose" their scale of growth or green lifestyle, it is as if the developed countries are supposed to take it as their mission to "help out" the less developed countries by imposing their advanced technologies and streamlining their financial resources (with the matter of sustainability funnelled down to the matter of measuring and controlling carbon footprints).

Secondly, while the sustainable development discourse embodies the contradiction of conjuring the scale of totality and aspirations to become global, in reality practitioners of sustainability have a limited capacity to handle only localized problems. While the terms of the discourse were clearly set by the scientists of developed countries that played a major role in polluting the earth, a significant disparity remained in the distribution of information surrounding the global debate on climate change (Chakrabarty 2017). As a result, the unevenly distributed discourse of sustainable development perpetuated the patronizing view of the West, even when it was disposed against the crisis generated by the West. When the representatives from developing countries are mobilized to engage with the global forums such as the UN Sustainable Development Solutions Network (SDSN) or the American National Standards Institute's Network on Smart and Sustainable Cities (ANSSC), it goes unquestioned how, in the respective countries, the issue of climate change and sustainable development is approached with varying degrees of urgency. These networks, designed to promote and transfer advanced technology, standards, and expertise from the developed countries to the rest of the world, embrace the moral responsibility to lead the rest of the world towards the common goal of reducing greenhouse gas emissions and substituting with renewable energy.

Both for the smart city and sustainable development discourses, science and technology prevail as an objective and value-neutral source of knowledge that can set up a universal standard to which countries and cities are supposed to submit themselves voluntarily. Herein the role of the cities and their governments is redefined through engagement with the technologically advanced, sustainable, and efficient ways of management, by imposing regulations and disciplines on themselves. A newly suggested mode of governing cities entails replacing the bureaucratic management with automated technology, scientific standards, and protocols embedded throughout a number of receptacles in the city, while the residual forms of governing entities are appointed to a seemingly reduced role of steering the machinic city.

Conclusion

Both sustainable development and the smart city, as a discourse and a programme of development and governance, share an aligned interest, goal, and approach to distinguish themselves from the previous model of development and governance. Their perceived problems are global in scale, and the solutions are technical. The problem is that the offered solutions only scratch the surface of the existing depth and complexity of the problem if they have anything at all to do with actually solving the problem in the first place.

Moreover, the sterile vision advanced by smart city-enabled sustainable development, according to one of David Harvey's interviews with Evgeny Morozov, is not quite convincing because it merely suggests a "technological fix to the problems of social relations in the city" (Harvey 2016). Problems in cities are more than just regulating traffic, measuring air quality, and managing sewage systems. They are problems of social relations. Environmental science and technology establish a problem that implicates the whole of humankind, which is defined, diagnosed, treated, and resolved by data. This conversion of decisions regarding social relations into an algorithmic process not only diminishes the space for citizen engagement but displaces political accountability itself, which recedes to the supposedly value-neutral scientific and technological standards. Especially when such transition is justified by ethical and global imperatives, it is even harder to challenge its rationale.

To summarize, this chapter traced a historical evolution of sustainable development discourse and one of its articulations in the technical urban design scheme of a smart city. Through a case study of Songdo, South Korea, the nexus between the discursive and material formations was explored. The exploration involved identifying the network of connections among the emergence of a universal techno-scientific standard, the datafication of the environment, the global imperatives targeting the cities, an alternative accumulation strategy, and the formation and idealization of new civic virtues and responsibilities for South Korean citizens in the twenty-first century.

The seemingly disjointed themes of sustainable development and the smart city successfully came together by sharing a common goal of solving the humanitarian crisis of planetary urbanization and climate change and by appointing technology as the primary solution to the problem. Cities have become increasingly entangled in the zone of global responsibilization of ethical and alternative modes of development, and the citizens have been called out as responsible subjects who consciously stylize their daily conduct to conform to the norm: to recycle, to exercise, to take care of oneself, and to participate in the universal challenge against climate change. The people, in this case, conceptualized as global humanity, are called into being as a political subject, thereby disavowing the radical heterogeneity and antagonisms that cut through "the people" (Swyngedouw 2010, p. 223).

Songdo is an exemplary case among many other cities in the world that are similarly situated at the intersections of this emerging mode of governing cities and the environment, which entails the displacement of politics by datafication, which is wedded to the formation of new citizenry ethics and the universal reward of going "green." It aligns with the goal of the advanced liberal government that relegates its responsibility to the supposedly value-neutral mechanism of data science and technology. In light of this finding, careful attention is needed to examine the process through which the entities that belong to the traditional category of political science are bypassed by the constant conjuring of global scale, and to theorize the hybridized assemblage of relationships into which economics and ethics, technology, and politics are brought together.

References

Albino, V., Berardi, U., & Dangelico, R. M. (2015). Smart cities: Definitions, dimensions, performance, and initiatives. *Journal of Urban Technology, 22*(1), 3–21.

Bhatt, V. (2016, December 21). U.S. Green Building Council Leadership in Energy and Environmental Design for Cities Program. American National Standards Institute Network on Smart and Sustainable Cities webinar.

Chakrabarty, D. (2017). The politics of climate change is more than the politics of capitalism. *Theory, Culture & Society, 34*(2–3), 25–37.

Cisco. (2010). *Smart+connected communities: Changing a city, a country, the world.* https://www.cisco.com/c/dam/en_us/solutions/industries/docs/scc/09CS2326_SCC_BrochureForWest_r3_112409.pdf. Accessed 10 Oct 2018.

Eger, J. M. (2005). Smart communities, universities, and globalization: Educating the workforce for tomorrow's economy. *Metropolitan Universities, 16*(4), 28–38.

Firger, J. (2017, December 15). Can tech save the planet? Cisco thinks so. *Newsweek.* https://www.newsweek.com/green-rankings-2017-cisco-systems-iot-tech-sustainability-green-rankings-2017-18. Accessed 10 Oct 2018.

Gabrys, J. (2013). *Digital rubbish: A natural history of electronics.* Ann Arbor: University of Michigan Press.

Haque, M. S. (2006). Limits of "sustainable development" as a concept. In G. M. Mudacumura, D. Mebratu, & M. S. Haque (Eds.), *Sustainable development policy and administration.* Boca Raton: Taylor & Francis.

Harvey, D. (2016, November 18). *Conversation between David Harvey & Evgeny Morozov on post-neoliberalism, Trump, infrastructure, sharing economy, smart city.* Video on blog, Reading Marx's *Capital* with David Harvey. http://david-harvey.org/2016/11/video-conversation-between-david-harvey-evgeny-morozov-on-post-neoliberalism-trump-infrastructure-sharing-economy-smart-city/. Accessed 11 May 2017.

James, I. (2016, October 14). *Songdo: No Man's City*. Korea Exposé. https://koreaexpose.com/songdo-no-mans-city. Accessed 11 May 2017.

Kim, C. (2010). Place promotion and symbolic characterization of New Songdo City, South Korea. *Cities, 27*, 13–19.

Koopman, C. (2013). *Genealogy as critique: Foucault and the problems of modernity*. Bloomington/Indianapolis: Indiana University Press.

Latour, B. (1998). To modernize or to ecologize? That's the question. In N. Castree & B. Willems-Braun (Eds.), *Remaking reality: Nature at the millennium* (pp. 221–242). London/New York: Routledge.

Lee, K. (2002). *Toward a cultural history in the Korean present: Locating the cultural politics of the everyday*. PhD thesis, University of Illinois at Urbana-Champaign, Urbana-Champaign.

Maxwell, R., & Miller, T. (2012). *Greening the media*. New York: Oxford University Press.

Osborne, T., & Rose, N. (1999). Governing cities: Notes on the spatialisation of virtue. *Environment and Planning D: Society and Space, 17*, 737–760.

Park, B. G. (2005). Spatially selective liberalization and graduated sovereignty: Politics of neo-liberalism and "special economic zones" in South Korea. *Political Geography, 24*, 850–873.

Park, Y. S. (2008). *A man who changed the map of Korea*. Seoul: Hanguk Gyeongje Shinmun. (in Korean).

Redclift, M. (2005). Sustainable development (1987–2005): An oxymoron comes of age. *Sustainable Development, 13*, 212–227.

Songdo IBD. (2013, August 23). *Reducing energy consumption: Not so difficult with Songdo International City! (Korean)*. Songdo International Business District official blog. Gale International Korea. http://songdoibd.tistory.com/507. Accessed 25 May 2017.

Swyngedouw, E. (2010). Apocalypse forever?: Post-political populism and the spectre of climate change. *Theory, Culture & Society, 27*(2–3), 213–232.

Udall, R., & Schendler, A. (2005, August 9). *LEED is broken – Let's fix it*. iGreenBuild.com, http://www.igreenbuild.com/cd_1706.aspx. Accessed 10 May 2017.

Vanolo, A. (2014). Smartmentality: The smart city as disciplinary strategy. *Urban Studies, 51*(5), 883–898.

Wilson, D. (2007). City transformation and the global trope: Indianapolis and Cleveland. *Globalizations, 4*(1), 29–44.

World Commission on Environment and Development (WCED). (1987). *Our common future*. United Nations report. http://www.un-documents.net/our-common-future.pdf. Accessed 11 May 2017.

Yi, W. W. (2012, October 15). Songdo striving to showcase 'Green City'. *The Korea Times*. http://koreatimes.co.kr/www/news/nation/2012/10/113_122326.html. Accessed 11 May 2017.

Yoneda, Y. (2010, April 13). Songdo Convensia is First LEED-Certified Convention Center in Asia. *Inhabitat*. http://inhabitat.com/songdo-convensia-is-first-leed-certified-convention-center-in-asia. Accessed 11 May 2017.

Sparking an Interest in Local Water Issues: Social Media as an Enabler for (Hyper)local Community Engagement in Environmental Issues and the Role of Solution Journalism

Franzisca Weder

Global and Local Communication of Water Issues

According to the United Nations, around two billion people, which equals one-fifth of the world's population, live in areas of water scarcity. Another 1.6 billion face economic water shortage (UN 2018). In total, two-thirds of the world's population experience water scarcity for at least one month a year (Mekonnen and Hoekstra 2016). A third of the biggest groundwater systems of the world are already in distress (Richey et al. 2015) and by 2025, not enough (potable) water will be available for two-thirds of the world's population (UN-Water and FAO 2007, pp. 9–10), with Asia as the most affected region (Burek et al. 2016).

F. Weder (✉)
School of Communication and Arts, The University of Queensland, Brisbane, QLD, Australia

J. Díaz-Pont et al. (eds.), *The Local and the Digital in Environmental Communication*, Global Transformations in Media and Communication Research - A Palgrave and IAMCR Series,
https://doi.org/10.1007/978-3-030-37330-6_11

To meet this challenge and ensure availability and sustainable management of water and sanitation for all in the future (SDG #6, UN 2017), "fundamental changes in water management are necessary" (Rosegrant et al. 2010, p. 17). Therefore, the UN proposes "integrated water resource management" as a "broad framework for governments" and regional adaptions (UN 2018). Even in Europe and particularly in water-rich countries like Austria, Slovenia or the southern part of Germany, where tensions regarding water issues are rising (privatization of water, droughts, etc.), sustainable solutions for water management in the future are needed. However, like climate change, the challenges around water supply and management often feature invisible symptoms with high complexity, which seem to go along with an invisible, non- or barely existing public discourse about sustainable water management (Voci and Weder 2015). While the value of increased stakeholder involvement in national river basin management planning has been recognized in recent years (EU 2018; MDBA 2018; Kaika 2003), there is a lack of in-depth research into the communication processes and factors that support fruitful awareness, discourses and—particularly—involvement on a local level (Tippett et al. 2016).

Mass media are accountable for what and how they communicate about environmental issues (Krcmar et al. 2016). Furthermore, mediated debates—especially on a local level—influence and shape environmental knowledge (Lester 2010, p. 181, 183), construct the public's perception of the environment (Cox 2013; Anderson 1997), and enable or hinder political engagement (Christians et al. 2009; Hackett 2017).

Today, social media like Facebook, YouTube or Twitter have an impact on the role of media in general and on issue lifecycles and public discourses (Van Dijck and Poell 2013). The growth of easy-to-use digital publishing technologies and the related audience behaviour influence public communication and the creation, as well as the configuration of public spaces. Considering the dynamics of mediatization and digitalization, it can be stated that we have reached an era of "networked journalism", with a concept of journalists working with the participation and engagement of the public (Kröll 2015). Additionally, new "participatory" communicator roles (citizen journalists, blogger, etc.) (Wall 2017) challenge our established principles and practices of journalism (Neuberger 2014; Lee-Wright et al. 2012; Harcup 2004; Pavlik 2001). Thus, in rethinking public discourses around environmental issues like climate change, degrading ecosystems or resources facing the risk of scarcity like water, the question arises as to how the information about potential risks and long-term solutions can be presented. Furthermore, we ask the following questions: how do people get engaged in those issues? More precisely: how to lure the rather passive and "not-in-my-backyard"—or NIMBYs—out of their backyard? Lastly: what is the role of journalists in creating and configuring digital communication spaces on a local level?

The empirical evidence gained in case studies with narrative interviews in water-rich and water-scarce communities will be presented as the core of this chapter. The findings show that a (hyper)local digital communication space can help to stimulate and facilitate a public discourse on environmental issues and communicative participation. However, the interviews conducted in those communities show that how they experience water scarcity, as well as how this communication space is created and maintained, highly depends on the self-perception of the local journalists, their willingness to use social media and their actual role in the local digital communication space. Thus, after explaining the case studies and the interview method and presenting the findings, the chapter will end with a call for a reconceptualization of local environmental journalism as solution journalism.

Theoretical Background

The research area of environmental communication is fairly established (Hansen and Cox 2015; Cox 2013; Lester and Hutchins 2013; Lester et al. 2010). Many scholars debate the intersections of this field of research with science communication (Akin and Scheufele 2017; Hall Jamieson et al. 2017), risk communication (Lundgren and McMakin 2013) and media and communication studies in general (Cox and Pezzullo 2015; Bonfadelli 2010). Environmental communication research reveals high complexity of issues; overuse of buzzwords like sustainability by politicians and corporations, the general politicization of climate change and related issues (Pepermans and Maeseele 2016) and expert lay gaps as core problems (Hall Jamieson et al. 2017; Schiele et al. 2012). In the literature, a research gap can be identified concerning the role of local news media for community engagement: local issues are still newsworthy because of the factor "closeness" (Harcup and O'Neill 2016) and their linkage to the direct surrounding and sociopolitical context of individuals (Aldridge 2007; Miller et al. 2012). Even in today's digital age, the local community is still described as the most important and primary communication space

(Firmstone and Coleman 2015) or "field of discourse" (Weder 2018), with a "lack of glamour" but desperately wanted by the audience (Aldridge 2007). Despite the ongoing high interest in local news and issues of immediate concern in the era of convergence (Diehl and Karmasin 2013) or "webvergence" (Thorton and Keith 2009), local newspapers and broadcasters face economic pressure (Sasseen et al. 2013; Lin and Jeffres 2001). Here, complementary online engagement with their audiences not only influences their bottom line but offers new ways of reporting and storytelling, in particular, on a local level (Smith et al. 2007). Apart from single studies, referring to the local context (i.e. Ward and Van Vuuren 2013), this is not recognized in environmental communication research so far.

Generally, there are only limited definitions of "the local". It is defined as a sociocultural space that people define as their "homeland" with a high degree of identification. From a political perspective, a region or local community is a political unit. This local public sphere is mostly demarcated by the distribution area of a local or regional newspaper. Today, this local field of discourse is created not only in a local editorial office; especially in the era of transition to digital content and changed ideas of the term "public sphere" (Benson 2009; Habermas 1984), there are also smaller newsrooms (Moon and Hadley 2014) and journalists working in a cross-media setting (Diehl and Karmasin 2013). This means storytelling across media and across platforms, transmedia storytelling (Veglis 2012; Scolari 2009; Perryman 2008), increasing visual orientation and new forms of citizen journalism and production in general (Bruns 2008). Even though the printed versions of the newspapers are still considered as the backbone of local journalism, including more source diversity (Watts and Maddison 2014), they still struggle in the era of cross media and the age of free, digital information. Consequently, (hyper)local journalism is debated as an alternative (Downman and Murray 2018) to reporting events and cover topics which are on a very local scale. This includes websites or social media platforms of a specific neighbourhood or even a street or part of a community, which goes hand in hand with concepts of slow or small-scale journalism (Greenberg 2013). The common ground of these concepts is that only very local or less contextualized issues have more impact and stronger interactions with the audience than wider interpretations of issues.

Constantly transforming audiences (Carpentier et al. 2014) demand new strategies to bring readers directly into the conversation. Here, using social media is seen as an enabler of community participation (Meyer and Tang

2015; Lai and Tang 2015) and, therefore, for local democracy (Firmstone 2016; Chen et al. 2015; Barnett and Townend 2014). The literature on stakeholder engagement complements this assumption (O'Riordan and Fairbrass 2014; Greenwood 2007), defining community engagement as involvement in political decision-making processes (Crane et al. 2008) and pointing out the major role of digital engagement (e.g. O'Brian and Cairns 2016). Bringing both aspects together, the following research question served as a guideline for the case studies, presented in the following chapter:

[RQ]: What is the role and the impact of local newspapers and their social media performance on engaging the local community in environmental issues such as sustainable water usage?

Methodological Approach

The ten case studies explained in the following section were conducted to learn about the role of local newspapers and their social media performance, as well as the role of journalists in creating and facilitating digital communication spaces. This space is seen as an enabler of individual and collective engagement in environmental issues such as sustainable water usage on a local or community level. The cases were chosen around river basin and water resource management policy or actual projects with a high local impact and existing examples of local engagement, social movements or resistance as response to changing environmental scenarios. The case studies were conducted in water-rich (Europe: Danube/Austria, Germany, Serbia and Romania; Austria: River Mur) and water-scarce regions (Australia: Murray River; South Africa: Water reservoirs around Cape Town; USA: San Bernardino County, CAL) (see Table 1).

In all cases, river basin and resource management plans are mostly created and executed on a (trans)national level (European Union; Commonwealth), including various regulative processes, consultations, political and communication strategies around building dams, weirs, as well as related interventions into nature on a local level. The cases are different related to their degree of institutionalization of engagement, involvement of various stakeholder groups and their very specific timespan. Thus, we had to deal with only limited numbers of interviewees in some cases which worked more as "expert interviews" and reference to the case studies at the end. However, in all cases on all levels (transnational, national, regional and local), the design of the study, from the first inves-

Table 1 Case studies and number of interviews

	Transnational		*National/regional*			*Local*				
CASES CATEG.	*Europe, Danube*	*Europe, Alpe Water Scarce*	*South Africa, Cape Town*	*Australia, Murray River*	*New Zealand, Waikato River*	*Germany, Salzach*	*Austria, Graz/"Save the Mur"*	*Austria, Klagenfurt/ citizen initiative*	*Austria, Kärnten wasser. reich*	*US, San Bernardino*
POL	8	1		3		2				
NGO/ ACT	3		3	4		1	3	2	3	
ECON	2		1	6				1		1
FARM			1	3				2		
LAY	2		2	9	1					
SCI		2		3	1					
JOURN	1		1	4				1		
PR			1	2		1		1		
Total	**16**	**3**	**9**	**34**	**2**	**3**	**3**	**7**	**3**	**1**
SUM	**81**									

Fig. 1 Local environmental movement as response to water management plans. (Photo: Private)

tigation and research to the interviews, was to start on a hyperlocal level. Rather than analysing stakeholder-dialogues planned by the national ministry and "implemented" top down, for CASE 8 ("Rettet die Mur"/"Save the Mur"), we started to walk along the River Mur in Graz/Austria, to talk to people who were tying themselves on trees that were meant to be cut down for a new weir and to analyse the discourse on the Facebook page "Save the Mur",[1] the platform of the social movement fighting against the dyking of the watercourse and infringement into the natural flow of the river (Fig. 1).

Case studies are perceived as the means of choice when an in-depth understanding of an entity, as it is situated in context, is needed (Bryman 2016; Yin 2009; Creswell 2007). Thus, starting with CASE 1 ("Donau"/"Danube"), we chose a snowball-system in finding partners

[1] https://www.facebook.com/rettetdiemur.at/

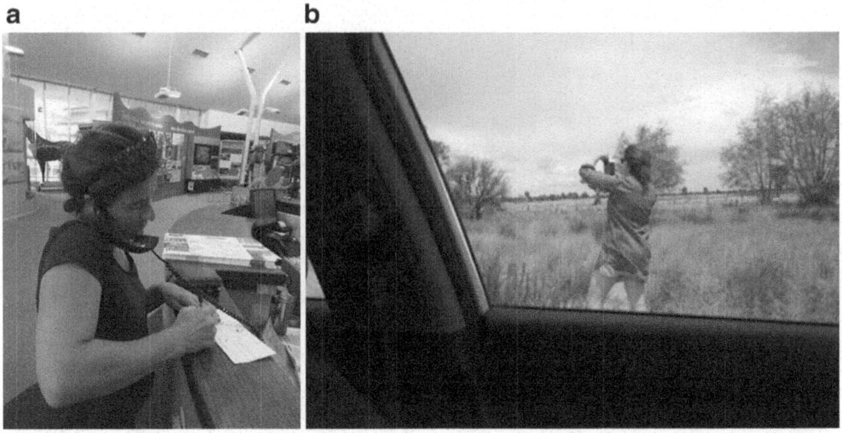

Fig. 2 (**a**, **b**) "Critical snowballing" along Murray River; interviews, documentation and experiencing a water-scarce country. (Photo: Private)

for narrative interviews, where we always asked one interview partner about their networks and for other potential interviewees they might know of, dealing with water (water management, water politics, environmental issues, activism, etc.) (Brodschöll 2003; Weder 2017). The process of "critical snowballing" in finding interview partners was very successful in CASE 2 ("Murray River") by conducting a field trip along the Murray River Basin in Australia (see Fig. 2a, b).

Generally, interviewing is one of the most intimate and rewarding communication research methodologies in environmental communication studies. "It is necessary to listen not only with the tidiest and most precise of one's cognitive abilities", Grant McCracken explains in *The Long Interview* (1988), "but also with the whole of one's experience and imagination" (p. 19). Thus, with an explorative and critical research approach, room for the individual experience of the interviewer was left. By the same time, we chose the cases related to their variety of related communication processes. Therefore, although we had nine cases already, we decided to do another case study with interviews in Cape Town, South Africa, in 2018, just before the proclaimed and highly mediatized "day zero" (2018)[2] was "scheduled" (see activists in action in Fig. 3).

[2] http://coct.co/water-dashboard/; after the official day zero, it was postponed for another year by the government. This can be reflected with various theoretical approaches, which would exceed this chapter.

Fig. 3 Activists in Cape Town, April 2018, shortly before (the first announcement of) "day zero". (Photo: Private)

In total, we were able to conduct 81 narrative interviews related to the case studies. The text corpus of the interviews was analysed with a qualitative text analysis developed by Philipp Mayring (Franzosi 2007; Mayring 2000, 2002). This type of summarizing content analysis starts with an explicit definition of the most important claims and arguments in the material related to the research question; this empirically grounded process relates the interviewee and the interviewer to the context of the interview. The analysis of the material is then done in two phases of reduction, where similar phrases are bundled and summarized. By doing so, the material gets generalized and categorized, which means it can be analysed on a higher level of abstraction at the end, connections between the interviews as well as issue clusters can be built. The main reason to work with this type of qualitative content analysis was its fit to the "critical snowballing" type of interviewing and its adjustable question-response setting, which could reveal the importance of the interviewees' subjective perceptions of the water issue, the media and their individual role (Edwards and Holland 2013, pp. 77–78; Denzin and Lincoln 2007; Douglas 1970).

The interviews were held between January 2013 and May 2018 due to the specific cases and event-related environmental movements and forms of communicative action and activism mentioned above. Related to the aforementioned research question, the basic assumption for the case studies was that there is a changing role of the local communication spaces and challenges of digital media on the community level as well as a changing role of journalists and other communicators and their cross media work in enabling citizens to engage in local issues and in local processes of deliberation and democracy (Firmstone and Coleman 2014; Hackett et al. 2017). Therefore, politicians [Pol], NGO people [NGO], farmers and irrigators [Farm], activists [Act], scientists [Sci] and laypeople [Lay], but mainly communication strategists and PR professional [PR] and journalists [Journ] from local newspapers were interviewed.

Findings

All interviewed journalists and communicators related to a local newspaper point out that they face the "effects" of digitalization mentioned in the theoretical section: firstly, an increasing influence of PR and marketing and related challenges like the development from a "logic of journalism" towards a source-driven reportage of science (Bauer and Gregory 2007). Secondly, an ongoing fight for attention in an era of scandalizations and decrease of trust in media in general. As well, nearly all interviewees mention the "green fatigue", as described in the literature (Lester 2010, p. 181), which could be mainly traced back to the increasingly plural science media ecosystem (Fahy and Nisbet 2011) with more arguments, perspectives and polarization, with new contexts and contextual clues. This seems to be even worse in science and environmental communication and, in particular, in dealing with the issue of water supply as rather complex issue, debated from a political perspective (privatization of water, regulations, river basin management, security), an economic perspective (hydropower, water management, privatization, resource management) and an environmental perspective (ecosystems, wet lands and natural reservoirs, pollution, etc.). The findings in detail:

Environmental Issues Are (Hyper)local

The literature states that environmental issues are mostly (hyper)local issues (Dernbach and Heuer 2000). The interviewed journalists from local newspapers point out that they focus on news that would usually not be covered by larger mainstream media outlets (IJour2/Murray), which is confirmed by media and communication scholars that have mostly studied local media, social media or environmental journalism (ISci10/Germany). Furthermore, the interviews show that in the newspaper's Facebook communication efforts, the local news are taken to the extreme: "for national newspapers, a new weir or dying trees are just a splash, whereas for our local papers it's an ongoing story" (IAct/Murray). Requests to engage happen via social media complements. "It's the flavour of the month"; "it's not so much in the media, it's at the fringes, in social media and academic worlds" (ILay/Murray); "I would like to see more people call in and say, what they just experienced" (IJour2/Murray). Social media enable new forms of mobilization and participation (Enjolras et al. 2013). Thus, new, as well as traditional media, not only play a vital role for the articulation of stakeholder expectations regarding resistance and activism but they also communicate the allocation of responsibility to specific sectors, corporations and even political institutions and related actors. The journalists describe their main function in a local setting as reporting, consulting and negotiation processes, informing people and being a mouthpiece: "We also act as communicators for environmental groups"; "when we write stories, we try to make sure that the major stakeholders' voices are represented" (IJour4/Murray). This is undermined by other journalists, that is, in South Africa: "You need diversity in views, that makes people want to debate as well" (IJour/SA).

New Potential for Environmental Communication in Local Social Media

Generally, local journalists are judged by NGO people, farmers and laypersons as having a "good connection to the local people" and "competence in dealing with us locals" (ILay/SA). Moreover, it is mentioned, that local journalists can stick with an issue for much longer compared to national ones (IJour/A). Thus, the barriers to engage the local community, mostly seen in their complementary social media activities—namely Facebook—were described as "issue-generator" and "catalyst" (IJour3/Murray).

Overall, the findings suggest a new role of local media and their social media activities in dealing with environmental issues. Social media helps to re-establish a relationship with the community and stimulates particularly passive people to step out of their backyard and take a voice (Hermans et al. 2014) because social media relates people to each other and enables connectivity between people as well as between people and issues.

Never Underestimate the Letter to the Editor

Generally, young people are not engaged with the local newspaper, but "they call in or comment on Facebook, asking can you write a story" or they start a petition against a political decision online (IJour2/Murray; IExp/Brussels; ILay/Serbia). Even though social media activities of local newspapers are highly valued, the letter to the editor is still very important: the "old fashioned one", directed to the print version of the newspaper. However, even in the digital age, "we have a lot of letters to the editor, a lot of texts to the editor", even "on Facebook it is directly addressed to the editor" (IJour3/Murray). This is recognized by the newspaper: The editor of the same newspaper in South Australia responds: "I do interact with them quite regularly, we've been doing a lot of work with them [with the authors of the letters to the editor]" (IJour5/Murray). This is highly valued by other communicators in the same region: "I think you need to build some strong personal relationships across the community to the government ... there need to be very strong interpersonal relationships" (IFarm/Murray). Talking about environmental issues, the potential of Facebook for local newspapers is explicitly mentioned as new version of the letter to the editor and communication hub mainly in sparsely inhabited regions.

Journalists as Facilitators: Rebels Wanted!

Of course, it takes time to deconstruct topics like river basin management plans, water power stations or drought prevention strategies. Thus, "slow journalism" is debated as one concept to lure the NIMBYs out of their fortress in the literature (Le Masurier 2016; Greenberg 2013). Indeed, the dead fish in the wetlands or the trees that have to be cut for the new weir get reflected more strongly in the local paper. The journalists claim that they need strong interpersonal relationships with the inactive people out there and "love to see people call in and say, what they just experienced—but they don't do it related to water" (IJour2/Murray). Here, journalists perceive themselves as facilitators of a discourse: "I'm in the job, I'm in an industry where my discourse is happening around problematizing things" (IJour1/Murray). The self-perception as a "rebel" does not come up in the interviews: "Government doesn't necessarily like to use us as a voice, because it tends to be a bit confrontational when they can't control the way they look, ... I'm sure that they try and get the message out through direct communication and information sessions" (IJour4/Murray). The perspective of activists and NGO communicators is quite different: "We need journalists who uncover the hidden networks of water management" (IAct/SA); "Only in social media people blame everything or everyone. Mainly the government" (IFarm/Murray). While activists smell trouble and talk about conspiracies, local journalists seem to be more passive: "Sometimes the environment itself just becomes so problematic that you cannot ignore the fact that there is no water because it's not raining. And so then it's not the government that sources the problem, it's just the environment you live in" (IJour2/Murray). Evidently, journalists perceive themselves as constructivists in facilitating discourses—without steering them or particularly initiating them.

Concluding Remarks

In this chapter, the following overarching research question guided the empirical study: What is the impact of local newspapers and their social media performance on engaging the local community in environmental issues such as sustainable water usage? The answer to the first question is not simple. Only the interviews with activists and NGO people showed that the issue of water supply, management and the risk of scarcity is (purposely) not reported in the media; community networks are active, "it's all about politics" (IAct/SA). "Everything bad happening related to Day Zero[3] is related to political changes and power issues" (IEnvAct/SA); "there is always a conspiracy" (ILay/SA), therefore, "the communities themselves have their own networks of groups and they meet face to face and they talk on social media" (IJour4/Murray).

[3] https://www.reuters.com/article/us-safrica-drought/cape-town-day-zero-pushed-back-to-2019-as-dams-fill-up-in-south-africa-idUSKCN1HA1LN, retrieved on July 10, 2018.

Local journalists seem to play an important role in creating a digital communication space via their social media activities and the newspaper's Facebook page. They perceive themselves as an "interactive facilitator": "we try to come to some common understanding of water related issues" (IJour4/Murray); "we develop, I guess, some mutually acceptable processes to go through, so we do those things a bit more closely together" (IAct/Murray). "I go to all the consultations; I am there if local people meet" (IJour2/Murray). However, the individual engagement of a journalist depends on the community—rather than the issue.

To sum it up, social media on a local level seem to be a double-edged sword. When people are passionate enough about something to do research and find out about it, they are the ones that bring the most engaging attitudes, opinions and inputs. Rather than the people who will just jump on Facebook and comment: "that's stupid" (IJour2/Murray). However, putting an issue on the local agenda is only one side. It is easy to click on something and like it, or to sign an online petition, but that is not the same as actually "doing something" (IAct/SA). Thus, social media engagement of local newspapers seems a necessary and useful complement for community engagement—but not suitable for awareness and long-lasting, sustainable involvement with environmental issues in general. The readership that relates themselves to environmental issues is rated as quite small and they are active anyway, with or without the local newspaper talking about those issues. Others are more engaged in the "juicy sort of gossip kind of stuff, like the who got arrested on the weekend" (IJour5/Murray).

Discussion and Theoretical Conceptualization of the Findings

Water resources and supply, water politics and water management are sensitive issues; reporting the issues is a multifaceted challenge for local newspapers. Individual behavioural change is influenced not by rational information or logical processes, but by emotions, changes in social norms and the behaviours of others. Here, social media—not in general but local newspaper pages—and local online newsfeeds can play a role in triggering a public debate and bringing up an issue. In particular, national or global issues can be "linked" to the local environment. In this case, local newspapers can function as enablers of the discourses. Offline engagement such as events, opening up editorial meetings or other opportunities for direct dialogue may help to be more engaging. Online content, as well as further engagement and interactivity with the readers via Facebook, seems to be an option to overcome scepticism, alleged conspiracies, general politicization or influence by PR agencies, mentioned in the interviews mainly by local journalists. However, journalists as well as PR professionals mentioned the role of local radio as something that should be considered for further research. For future studies in environmental communication, it can be stated that the influence of local print-journalists does not seem to be as high when compared to their colleagues from local radio stations: "ABC rural or ABC Riverland,[4] oh, there are different voices ... it's very important for environmental issues!" (IFarm/Murray). "ABC plays a very big role, particularly in regional areas, and those are particularly important because of the key stakeholders" (IAct/Murray). "Everyone listens to the radio" (ILay/Serbia), "Radio is on ... the whole day. Regulations, restrictions in water use are constantly communicated—and we listen" (IJour/SA). This implies future research potential.

Looking at the numbers of interviews related to the case studies, there is an unequal proportion of the interviews in the benefit of Australia, regarding number of the interviews (see Table 1). By knowing this and by reflecting the selectivity of the interviewed communicators and the strong influence of different cultural backgrounds as the major limitation of this study, water management and sustainable supply seem to be broad issues with political, economic as well as environmental aspects. More and more journalists, supported by the interviewed NGO people, laypersons and consultants, realize that with their reporting on environmental issues on a local level and with engaging a conversation with people that have, so far, nothing to do with those issues, they personally contribute to sustainable development. Their stories show that environmental communication in today's public media needs a deeper reflection and critical journalism with a deeper understanding of ethics. The reflection on media ethics seems to be a hypodermic negotiation process in every kind of environmental communication.

Here, the concept of solution journalism is introduced. There is a bunch of literature, dealing with "constructive journalism", with visions (i.e. Gyldensted 2015), academic reflections (i.e. Krüger 2016), practical

[4] http://www.abc.net.au/radio/riverland/, http://www.abc.net.au/news/rural/, retrieved on July 10, 2018.

implications, institutions and guidelines (overview in Meier 2018). McIntyre (2015, p. 9) offers a definition of constructive journalism as an "emerging form of journalism that involves applying positive psychology techniques to news work in an effort to create more productive, engaging stories while holding true to journalism's core functions". This goes far beyond news reporting based on the old saying that "only bad news is good news"-logic. Environmental journalism, often driven by specific negative events (natural disasters, flooding, heatwaves, droughts, etc.), is competing with news on war, unemployment or terrorism (Cox 2013, p. 145; Harcup and O'Neill 2016), as well as arguments between climate scientists versus global warming sceptics, angry residents versus corporations and so on.

Apparently, solution journalism and local journalism find equal ground in this constructive perspective, which goes beyond the idea that conflict is the key criterion for news stories. Instead, journalism has a specific approach in telling the good stories and offering solutions for societal problems (McIntyre 2017). On a micro level, journalists become "first movers" (Maier 2018) and perceive themselves as change agents or even activists (Weder and Krainer 2018). This could work in particular for the phenomenon of sustainability issues such as water supply (Newig et al. 2013, p. 2978), which could be characterized by complexity, uncertainty and ambivalence and influenced by conflicts of interest and values, as well as contradictions around the implementation of sustainability-related measures (Voci and Weder 2015). The interviews conducted within the case studies show that the potential of digital communication spaces created and maintained on a local level is not used so far—but is seen as the future of environmental journalism. On the local level, environmental and social movements are created or redefined as response to changing environments (droughts) or new scenarios (water management and river basin management plans); "environmental conflict is often fought locally, and it is from here that solutions will inevitably emerge" (Lester 2010, p. 182).

With the results of the study presented in this chapter, the concept of solution journalism and positive, constructive perspectives on news reporting can be reflected with Hackett's roles of journalists (2017, p. 40 ff.), to conceptualize the potential of local journalists to create and maintain local digital communication spaces and involve the audience in environmental issues such as water supply. On a local level, journalism can be firstly seen in Hackett's "facilitative role", in terms of enabling a public discourse, fostering debates and arguments; particularly in the letters to the editor section and the social media complement of the print version. Furthermore, supported by Gunster (2017, p. 53), the "radical role" can be realized in a digitalized world. The idea of actively advocating and mobilizing and communicating *for* sustainability fits with the concept of solution journalism dealing with (hyper)local, small-scaled issues, willing to prioritize and experiment with interventions.

On a local level and regarding water supply as an issue of concern, the interviews of the study presented here showed that the "slow", facilitative role is prevalent; mainly in terms of promoting dialogue in particular on social media platforms. However, there is still unused potential of "solution journalism" and a more radical role in terms of offering positive scenarios of a sustainable future. By doing so, journalists could challenge the existing status quo radically. A radical role and an understanding of journalism to mobilize opinion against societal power structures (Foxwell-Norton 2017, p. 151; Christians et al. 2009, p. 181) are not mentioned in the interviews. Further studies should explore the stimuli that would be necessary for journalists to offer future scenarios more radically and solutions to existing, very specific problems on a local level.

References

Akin, H., & Scheufele, D. (2017). Overview of the science of science communication. In K. H. Jamieson, D. Kahan, & D. A. Scheufele (Eds.), *The Oxford handbook of the science of science communication* (pp. 25–33). Oxford: Oxford University Press.

Aldridge, M. (2007). *Understanding the local media*. Maidenhead: Open University Press.

Anderson, A. (1997). *Media, culture and the environment*. New Brunswick: Rutgers University Press.

Barnett, S., & Townend, J. (2014). Plurality, policy and the local: Can hyperlocals fill the gap? *Journalism Practice, 9*(3), 332–349. https://doi.org/10.1080/17512786.2014.943930.

Bauer, M., & Gregory, J. (2007). From journalism to corporate communication in post-war Britain. In M. Bauer & M. Bucchi (Eds.), *Science, journalism and society: Science communication between news and public relations* (pp. 33–52). London: Routledge.

Benson, R. (2009). Shaping the public sphere: Habermas and beyond. *American Sociologist, 40*, 175–197. https://doi.org/10.1007/s12108-009-9071-4.

Bonfadelli, H. (2010). Environmental sustainability as challenge for media and journalism. In M. Gross & H. Heinrichs (Eds.), *Environmental sociology: European perspectives and interdisciplinary challenges*. https://doi.org/10.1007/978-90-481-8730-0_15.

Brodschöll, P. C. (2003). *Negotiating Sustainability in the Media: Critical perspectives on the popularisation of environmental concerns.* Curtin University of Technology, Faculty of Media, Society and Culture.

Bruns, A. (2008). *Blogs, Wikipedia, second life, and beyond: From production to produsage.* New York et al.: Peter Lang.

Bryman, A. (2016). *Social research methods* (5th ed.). Oxford: Oxford University Press.

Burek, P., Satoh, Y., Fischer, G., Kahil, M. T., Scherzer, A., Tramberend, S., Nava, L. F., Wada, Y., et al. (2016). Water futures and solution – Fast track initiative (final report). *IIASA Working Paper.* IIASA, Laxenburg: WP-16-006.

Carpentier, N., Schrøder, K. C., & Hallett, L. (2014). Audience/society transformations. In N. Carpentier, K. C. Schrøder, & L. Hallett (Eds.), *Audience transformations: Shifting audience positions in late modernity* (pp. 1–14). New York: Routledge.

Chen, N., Ognyanova, K., Zhang, C., Wang, C., Ball-Rokeach, S., & Parks, M. (2015). Causing ripples in local power relations. *Journalism Studies, 18*, 710. https://doi.org/10.1080/1461670X.2015.1078738.

Christians, C., Glasser, T., McQuail, D., Nordenstreng, K., & White, R. A. (2009). *Normative theories of the media: Journalism in democratic societies.* Urbana/Chicago: University of Illinois Press.

Cox, R. (2013). *Environmental communication and the public sphere* (3rd ed.). Los Angeles: Sage.

Cox, R., & Pezzullo, P. C. (2015). *Environmental communication and the public sphere* (4th ed.). London: Sage.

Crane, A., McWilliams, A., Matten, D., Moon, J., & Siegel, D. S. (Eds.). (2008). *The Oxford handbook of corporate social responsibility.* New York: Oxford University Press.

Creswell, J. W. (2007). *Qualitative inquiry & research design. Choosing among five traditions* (2nd ed.). Thousand Oaks: Sage.

Denzin, N., & Lincoln, Y. (Eds.). (2007). *The Sage handbook of qualitative research.* Thousand Oaks: Sage.

Dernbach, B., & Heuer, H. (2000). *Umweltberichterstattung im Lokalen. Ein Praxishandbuch.* Wiesbaden: Westdeutscher Verlag.

Diehl, S., & Karmasin, M. (2013). *Media and convergence management.* Heidelberg u.a.: Springer.

Douglas, J. D. (1970). Deviance and respectability: The social construction of moral meanings. In J. D. Douglas (Ed.), *Deviance and Respectability. The Social construction of Moral Meanings* (pp. 3–30). New York: London.

Downman, S., & Murray, R. (2018). *Hyperlocal journalism and digital disruptions. The journalism change agents in Australia and New Zealand.* New York: Routledge.

Edwards, R., & Holland, J. (2013). *What is qualitative interviewing?* London: Bloomsbury.

Enjolras, B., Steen-Johnsen, K., & Wollebæk, D. (2013). Social media and mobilization to offline demonstrations – transcending participatory divides? *New Media and Society, 15*(6), 890–908.

EU. (2018). The EU Water Framework Directive – Integrated river basin management for Europe. Retrieved July 10, 2018, from http://ec.europa.eu/environment/water/water-framework/index_en.html.

Fahy, D., & Nisbet, M. C. (2011). The science journalist online: Shifting roles and emerging practices. *Journalism, 12*(7), 778–793. https://doi.org/10.1177/1464884911412697.

Firmstone, J. (2016). Mapping changes in local news. *Journalism Practice, 10*(7), 928–938. https://doi.org/10.1080/17512786.2016.1165136.

Firmstone, J., & Coleman, S. (2014). The changing role of the local news media in enabling citizens to engage in local democracies. *Journalism Practice, 8*(5), 596–606. https://doi.org/10.1080/17512786.2014.895516.

Firmstone, J., & Coleman, S. (2015). Rethinking local communicative spaces: Reflecting on the implications of digital media and citizen journalism for the role of local journalism in engaging citizens. In R. K. Nielsen (Ed.), *Local journalism: The decline of newspapers and the rise of digital media* (pp. 117–140). London: I.B. Tauris.

Foxwell-Norton, K. (2017). Australian independent news media and climate change reporting. The case of COP21. In R. A. Hackett, S. Forde, S. Gunster, & K. Foxwell-Norton (Eds.), *Journalism and climate crisis. Public engagement, media alternatives* (pp. 144–166). London/New York: Routledge.

Franzosi, R. (Ed.). (2007). *Content analysis. Benchmarks in research methods series.* Thousand Oaks: Sage.

Greenberg, S. (2013). Slow journalism in the digital fast lane. In R. L. Keeble & J. Tulloch (Eds.), *Global literary journalism: Exploring the journalistic imagination* (pp. 381–393). New York: Peter Lang.

Greenwood, M. (2007). Stakeholder engagement: Beyond the myth of corporate responsibility. *Journal of Business Ethics, 74*(4), 315–327.

Gunster, S. (2017). Engaging climate communication. Audiences, frames, values and norms. In R. A. Hackett, S. Forde, S. Gunster, & K. Foxwell-Norton (Eds.), *Journalism and climate crisis. Public engagement, media alternatives* (pp. 49–76). London/New York: Routledge.

Gyldensted, C. (2015). *From mirrors to movers. Five elements of positive psychology in constructive journalism.* Ggroup Publishing: Charleston, SC.

Habermas, J. (1984/[1981]). *The theory of communicative action, volume one. Reason and the rationalization of society* (Thomas A. McCarthy, Trans.). Boston: Beacon Press.

Hackett, R. A. (2017). Democracy, climate crisis and journalism. Normative touchstones. In R. A. Hackett, S. Forde, S. Gunster, & K. Foxwell-Norton

(Eds.), *Journalism and climate crisis. Public engagement, media alternatives* (pp. 20–48). London/New York: Routledge.
Hackett, R. A., Forde, S., Gunster, S., & Foxwell-Norton, K. (2017). Introduction. In R. A. Hackett, S. Forde, S. Gunster, & K. Foxwell-Norton (Eds.), *Journalism and climate crisis. Public engagement, media alternatives* (pp. 1–19). London/ New York: Routledge.
Hall Jamieson, K., Kahan, D., & Scheufele, D. A. (2017). *The Oxford handbook of the science of science communication*. Oxford: Oxford University Press.
Hansen, A., & Cox, R. (Eds.). (2015). *The Routledge handbook of environment and communication*. Oxford: Routledge.
Harcup, T. (2004). *Journalism principles & practice*. London: Sage.
Harcup, T., & O'Neill, D. (2016). What is news? News values revisited (again). *Journalism Studies, 18*, 1470. Retrieved from http://www.tandfonline.com/doi/pdf.10.1080/1461670X.2016.1150193.
Hermans, L., Schaap, G., & Bardoel, J. (2014). Re-establishing the relationship with the public: Regional journalism and citizens' involvement in the news. *Journalism Studies, 15*(5), 642–654. https://doi.org/10.1080/1461670X.2014.894373.
Kaika, M. (2003). The water framework directive: A new directive for a changing social, political and economic European framework. *European Planning Studies, 11*(3), 299–316. https://doi.org/10.1080/09654310303640.
Krcmar, M., Ewoldson, D. R., & Koerner, A. (2016). *Communication science theory and research: An advanced introduction*. New York: Routledge.
Kröll, A. (2015). The role of journalism in the digital age. Being a superhero or Clark Kent: Do journalists think that networked journalism is an appropriate tool to work with (in the future)? Reuters Institute Fellowship Paper, University of Oxford.
Krüger, U. (2016). Solutions journalism. In Deutscher Fachjournalisten-Verband (Ed.), *Journalistische Genres* (pp. 173–182). Konstanz: UVK.
Lai, C.-H., & Tang, T. (2015). Understanding local news consumption and community participation via the lens of information repertoires and media multiplexity. *Mass Communication & Society, 18*(3), 325–349. https://doi.org/10.1080/15205436.2014.995768.
Lee-Wright, P., Phillips, A., & Witschge, T. (2012). *Changing journalism*. Oxon: Routledge.
Le Masurier, M. (2016). *Slow Journalism, Journalism Practice, 10*(4), 439–447, https://doi.org/10.1080/17512786.2016.1139902.
Lester, L. (2010). *Media and environment. Conflict, politics and the news*. Cambridge/Malden: Polity Press.
Lester, L., & Hutchins, B. (2013). *Environmental conflict and the media*. New York: Peter Lang.
Lin, C., & Jeffres, L. (2001). Comparing distinctions and similarities across websites of newspapers, radio stations and television stations. *Journalism & Mass*

Communication Quarterly, 78(3), 555–573. https://doi.org/10.1177/1077699001078003l.

Lundgre, R.E., & Mcmakin, A.H. (2013). Risk Communication: A Handbook for Communicating Environmental, Safety, and Health Risks, Fifth Edition. New York: Wiley.

Mayring, P. (2000). Qualitative content analysis. *Forum: Qualitative Social Research, 1*(2), 20. Retrieved from http://www.qualitative-research.net/index.php/fqs/article/view/1089/2385. 14 Nov 2016.

Mayring, P. (2002). Qualitative content analysis – Research instrument or mode of interpretation? In M. Kiegelmann (Ed.), *The role of the researcher in qualitative psychology* (pp. 139–148). Tübingen: Huber.

McIntyre, K. E. (2015). Constructive journalism: The effects of positive emotions and solution information in news stories. Dissertation at the University of North Carolina at Chapel Hill.

McIntyre, K. E. (2017). Solutions Journalism. *Journalism Practice, 13*, 16. https://doi.org/10.1080/17512786.2017.1409647.

MDBA (Murray-Darling Basin Authority). (2018). *Developing the Basin Plan*. Retrieved July 10, 2018, from https://www.mdba.gov.au/basin-plan/developing-basin-plan

Meier, K. (2018). How does the audience respond to constructive journalism?: Two experiments with multifaceted results. *Journalism Practice, 12*, 764–780. https://doi.org/10.1080/17512786.2018.1470472.

Mekonnen, M. M., & Hoekstra, A. Y. (2016). Four billion people facing severe water scarcity. *Science Advances, 2*(2), e1500323. https://doi.org/10.1126/sciadv.1500323.

Meyer, K. M., & Tang, T. (2015). #SocialJournalism: Local news media on Twitter. *International Journal on Media Management, 17*(4), 241–257. https://doi.org/10.1080/14241277.2015.1107569.

Miller, C., Purcell, K., & Rosenstill, T. (2012). 72% of Americans follow local news closely. Pew Research Internet Project. Retrieved from http://www.pewinternet.org/2012/04/12/72-of-americans-follow-local-news-closely/

Moon, S. J., & Hadley, P. (2014). Routinizing a new technology in the newsroom: Twitter as a news source in mainstream media. *Journal of Broadcasting & Electronic Media, 58*(2), 289–305. https://doi.org/10.1080/08838151.2014.906435.

Neuberger, C. (2014). The journalistic quality of internet formats and services. *Digital Journalism, 2*(3), 419–433. https://doi.org/10.1080/21670811.2014.892742.

Newig, J., Schulz, D., Fischer, D., Hetze, K., Laws, N., Lüdecke, G., & Rieckmann, M. (2013). Communication regarding sustainability: Conceptual perspectives and exploration of societal subsystems. *Sustainability, 5*, 2976–2990. https://doi.org/10.3390/su5072976.

O'Brian, H., & Cairns, P. (2016). *Why engagement matters: Cross-disciplinary perspectives of user engagement in digital media*. Cham: Springer.

O'Riordan, L., & Fairbrass, J. (2014). Managing CSR stakeholder engagement: A new conceptual framework. *Journal of Business Ethics, 125*(1), 121–145.

Pavlik, J. V. (2001). *Journalism and new media*. New York: Columbia University Press.

Pepermans, Y., & Maeseele, P. (2016). The politicization of climate change: Problem or solution? *Wiley Interdisciplinary Reviews: Climate Change, 7*(4), 478–485.

Perryman, N. (2008). Doctor who and the convergence of media: A case study in 'transmedia storytelling'. *Convergence, 14*(1), 21–39.

Richey, A. S., Thomas, B. F., Lo, M. H., Reager, J. T., Famiglietti, J. S., Voss, K., Swenson, Y., & Rodell, M. (2015). Quantifying renewable groundwaterstress with GRACE. *Water Resources Research, 51*, 5217–5238. https://doi.org/10.1002/2015WR017349.

Rosegrant, M. W., Cline, S., & Valmonte-Santos, R. (2010). Global water and food security: Megatrends and emerging issues. In C. Ringler, A. K. Biswas, & S. Cline (Eds.), *Global change: Impacts on water and food security*. Heidelberg: Springer.

Sasseen, J., Olmstead, K., & Mitchell, A. (2013). Digital: As mobile grows rapidly, the pressures on news intensify. Pew Internet & American Life Project. Retrieved from http://stateofthemedia.org/2013/digital-as-mobile-grows-rapidly-the-pressures-on-news-intensify/

Schiele, B., Claessens, M., & Shi, S. (2012). *Science Communication in the World: Practices, Theories and Trends*. Dordrecht: Springer.

Scolari, C. (2009). Transmedia storytelling. Implicit consumers, narrative worlds, and branding in contemporary media production. *International Journal of Communication, 3*, 586–606.

Smith, L. K., Tanner, A. H., & Duhe, S. F. (2007). Convergence concerns in local television: Conflicting views from the newsroom. *Journal of Broadcasting & ElectronicMedia, 51*(4), 555–574. https://doi.org/10.1080/0883815070162.

Thorton, L., & Keith, S. M. (2009). From convergence to webvergence: Tracking the evolution of broadcast-print partnerships through lens of change theory. *Journalism & Mass Communication Quarterly, 86*(2), 257–276. https://doi.org/10.1177/107769900908600201.

Tippett, J., Searle, B., Pahl-Wostl, C., & Rees, Y. (2016). Social learning in public participation in river basin management—Early findings from HarmoniCOP European case studies. *Environmental Science & Policy, 8*(3), 287–299. https://doi.org/10.1016/j.envsci.2005.03.003.

UN. (2017). *Sustainable Development Goals*. 17 Goals to transform our world. https://sustainabledevelopment.un.org/sdg6. 2 Dec 2016.

UN. (2018). UN water, *Water Scarcity*. Retrieved July 8, 2018, from http://www.unwater.org/water-facts/scarcity/

UN-Water & FAO (Food and Agriculture Organization of the United Nations). (2007). *Coping with water scarcity*. Challenge of the twenty-first century. Retrieved from http://www.fao.org/nr/water/docs/scarcity.pdf

Van Dijck, J., & Poell, T. (2013). Understanding social media logic. *Media and Communication, 1*(1), 2–14. https://doi.org/10.12924/mac2013.01010002.

Veglis, A. (2012). From cross media to transmedia reporting in newspaper articles. *Publishing Research Quarterly, 28*(4), 313–324. https://doi.org/10.1007/s12109-012-9294-z.

Voci, D., & Weder, F. (2015). *Use and abuse of sustainability in the media: A comparison between sustainability rhetoric in Italian, Austrian and German newspapers*. Proceedings of the 3rd CSR COM, Ljubljana, pp. 114–124. Retrieved from http://csr-com.org/img/upload/proceedings_za_objavo.pdf

Wall, M. (2017). Mapping citizen and participatory journalism. *Journalism Practice, 11*(2–3), 134–141. https://doi.org/10.1080/17512786.2016.1245890.

Ward, S., & van Vuuren, K. (2013). Belonging to the rainbow region: place, local media, and the construction of civil and moral identities strategic to climate change adaptability, environmental communication. *A Journal of Nature and Culture, 7*(1), 63–79.

Watts, R., & Maddison, J. (2014). Print news uses more source diversity than does broadcast. *Newspaper Research Journal, 35*(3), 107–118.

Weder, F. (2017). Storytelling as a research method. In F. Weder (Ed.), *Problematization of water supply*, Final report. Klagenfurt: University of Klagenfurt.

Weder, F. (2018). Hollow notion of corporate social responsibility. Introduction and application of a frame field model to investigate CSR communication in public discourses in Europe. *Medien Journal, 1*(2018), 5–27.

Weder, F., & Krainer, L. (2018). *Reflexivity on Contradictions*. Self-definitions of PR-people, Journalists, V-/Blogger and Instagramer facing the Climate Crisis. Proceedings of BledCom 25, to be published.

Yin, R. K. (2009). *Case study research: Design and methods* (4th ed.). Thousand Oaks: Sage.

Lived, Local and Everyday: Understanding Media Use in Green Home Renovation

Aneta Podkalicka and Deb Anderson

Introduction

Issues of environmental risk, crisis and sustainability have prompted significant growth in cross-disciplinary scholarship that explores their politics, governance, economics, ethics and lived experience. While a wealth of research has identified the media (primarily news) as a key institution that can serve to amplify or attenuate public perceptions of such issues (Hansen and Cox 2015), the growing and distinctive field of environmental communication has shed light on shifting power structures in a diversifying media landscape (e.g. Koteyko et al. 2015). Recent research stretches

A. Podkalicka (✉)
Communications and Media Studies, School of Media, Film and Journalism, Monash University, Caulfield East, VIC, Australia
e-mail: aneta.podkalicka@monash.edu

D. Anderson
Journalism, School of Media, Film and Journalism, Monash University, Caulfield East, VIC, Australia
e-mail: deb.anderson@monash.edu

J. Díaz-Pont et al. (eds.), *The Local and the Digital in Environmental Communication*, Global Transformations in Media and Communication Research - A Palgrave and IAMCR Series,
https://doi.org/10.1007/978-3-030-37330-6_12

from analyses of the impacts of digital and social media on the practices of environmental not-for-profit organisations (Takahashi et al. 2015) to studies of peer-to-peer exchanges powered by the internet (Andersson and Öhman 2017; Arniani et al. 2014; Haider 2015), and from the greening of consumption (Connolly and Prothero 2008) to the political economy of environmental media (Brevini and Murdock 2017). At the same time, media researchers have begun to explore the intersections of the local and digital in relation to environmental sustainability, drawing on international examples of citizen journalism and urban grassroots activism, including the use of locative media (Podkalicka and Rennie 2018; also Rodgers 2018 on hyperlocal media). There remains much to learn, however, about how the media are experienced—as object, subject and agent—in the communication of environmental issues, notably in their lived, local and everyday cultural context.

This chapter takes the latter as its point of departure. Seeking a more nuanced, grounded understanding of how media and communication processes may impact upon and shape public understanding and decision-making in society, it engages with the *everydayness* of media practice and, more broadly, the significance of cultural context and thick texture of lived experience. A sign of the increasing maturity of the field of environmental communication has been its widening scope not only in terms of its focus but also of its methods of inquiry and conceptual frameworks. As Anders Hansen and Robert Cox (2015) write, the field has broadened in the past few decades to examine a vast range of media and communication forms and content, while engaging with an increasing diversity of theoretical and disciplinary traditions. Importantly, in research on the social implications of environmental communication, they detect 'an increasing sophistication and appreciation of the highly dynamic and complex ways in which environmental messages, images and beliefs are promoted, contested and circulated' (p. 7). Yet as Hansen elsewhere points out (2015), there remains a critical need to continue to redress an imbalance in relation to the field's traditionally media-centric focus. Recent studies of mediatisation have productively emphasised the consequences of digital media technologies on various spheres of human lives (see below), while research conducted by Australian scholars, notably Lester and Hutchins (e.g. 2009, 2012), has challenged "media-centricity" to consider the dynamics of the mediation process that determines, and the structural and institutional conditions that constrain, environmental communication. Another promising path, Hansen notes, lies in recent work on the "reframing" of climate change communication (e.g. Nisbet 2009; Nisbet and Kotcher 2009; Maibach et al. 2010)—studies that locate media within their lived context and, in effect, 're-emphasize the role played by peers and opinion-leaders in building public and personal awareness, opinions, and understanding with regard to controversial environmental issues' (Hansen 2015, p. 388).

We aim to build upon existing understandings of media influence on public perception by shedding light on the *situated* integration of media and communication in everyday life. Our intention is to reaffirm the distinctive value of cultural and media studies approaches to understanding environmental communication in its lived, local and everyday context—which therefore form an important part of the wider interdisciplinary field of environmental communication research.

This chapter uses an Australian study of situated media use and engagement in the context of home renovation to explore intersections of "the local" and "the digital", in this case concerning residential energy efficiency. Although the media occupy a significant place in the ways in which people renovate their homes, few studies have examined in a systematic and detailed manner the specificity of media use in this context. In relation to the broader topic of household energy efficiency, academic work has recognised the importance of social networks in adopting energy-efficient technologies in residential sectors, including the positive influence of local governments (Bale et al. 2013) and local networks (van der Schoor and Scholtens 2015). The examination of digital media for residential energy efficiency is also gathering pace. Marta Arniani and her colleagues (2014) have, for example, documented a set of initiatives that rely on digital platforms and participatory approaches to tackle social challenges including energy-use reduction, arguing for peer-to-peer collaboration and exchanges that digital media foster. Yolande Strengers and Larissa Nicholls have offered critical insights into the application of smart home technologies and their implications for energy consumption (Strengers and Nicholls 2017).

The Australian study homes in on the experience of renovation as narrated by renovators, tracing discourses of media presence in the everyday labour of home renovation through the lens of media practice—that is, considering 'media as the open set of practices relating to, or oriented around, media' (Couldry 2004, p. 117). In addition, the narratives are viewed in relation to ideas on "mediatisation", a concept in media studies that describes the meta-dynamics of 'media-related change' (Hepp 2013, p. 616) brought about by the pervasiveness of media and communication technologies. Insofar as contemporary home renovations are mediatised in nature, this particular concept proffers a useful means of scrutinising the dynamics of media use and engagement. As Jansson (2017, p. 3) states, mediatisation processes can be 'broken down into sub-processes', 'operationalised and studied empirically in particular time-space settings to illuminate mediatisation's multiple and contextually formed expressions'. Jansson argues thus (2017, p. 3):

> Mediatisation research should try, in a more elaborated way than has so far been the case, to unveil the inner tensions, ambiguities and contradictions of a society in which media technologies have become taken-for-granted parts of everyday life.

Indeed, scholars in cultural and media studies have long argued for the significance of "the everyday", or "the mundane"—those aspects of culture that perhaps most resist analysis and which may seem, as Meaghan Morris noted, 'just too banal to be of interest' (1990, p. 18). As Morris (1990) pointed out, banality itself can form both a mythic signifier and a mask for value judgement. In that light, importantly, the Australian home renovations study was specifically designed to add to research on the centrality of communication and construction of the environment by locating the use of traditional and digital media within the local, lived and everyday context of "mundane" domestic media consumption and production—even while home renovation could be regarded as a special period in household life (Hulse et al. 2015; Podkalicka et al. 2016). This reflected a commitment to account for the practical, communicative, material and ethical factors that have bearing on material consumption, while more broadly underscoring the value of taking a cultural approach to research on media, communication and environment.

The chapter first locates Australian household consumption within its broader social-environmental context, canvassing the challenges of research design and the value of taking a cultural approach to the topic. It then examines a raft of empirical material that clearly bears out the dynamics of media engagement (notably digital) as part of a broader suite of social practices anchored in local, lived, everyday experience. Here both digital and traditional media and "unmediated" social interactions and communication are found to affect local decisions and modes of consumption. The chapter concludes by outlining some of the implications of the complex interplay between the local and the digital for environmental communication research. Overall, the intention of the chapter is to reaffirm the distinctive value of cultural and media studies approaches to understanding environmental communication in its lived, local and everyday context—which forms an important part of the wider interdisciplinary field of environmental communication research.

A Cultural Approach to Australian Homes, Media and Energy Efficiency

Cultural context (e.g. Gibson et al. 2011; Podkalicka 2018) and everyday practices (Maller et al. 2012; Judson and Maller 2014) have been given prominence as key to understanding Australian household consumption in the environmental social sciences. Taking a cultural approach to this environmental topic seems especially relevant Down Under, where home ownership is associated with cultural identity—the 'Great Australian Dream' (Allon 2008, 2014), which is reportedly 'fading' due to problems in housing affordability (Sheppard et al. 2017)—and home renovation represents a $32 billion market (Housing Industry Association 2017; this is projected to grow to $34.71 billion by 2020). Domestic energy efficiency is one of the nation's big challenges. The majority of houses in the state of Victoria (where this chapter's authors reside), for example, 'were built before energy efficiency regulations were first introduced in 2005 and have a very low thermal efficiency' (Sustainability Victoria 2014, p. iii). In addition, the housing market in Australia is characterised by low-density building such as detached houses (although this is now changing in metropolitan cities such as Sydney, Melbourne and Brisbane; see McKeith 2016), with new dwellings on average built larger over the years, which has impacted on energy consumption. The ways in which houses are constructed as well as daily consumption practices account for household energy usage patterns. Household energy consumption research and public programmes focus then on house types and their thermal performance, and also energy consumption, including 'space conditioning, water heaters, cooking products and appliances' and fuel types (Harrington and Foster 2008, p. x) to better understand and support improvements in energy efficiency. Measures to improve efficiency—that is, 'using less energy to provide the same service' (Napp et al. 2012)—have been retrofits, upgrades and renovations of existing houses (for instance, insulation and double glazing) and the uptake of energy-efficient products and appliances, including renewables (Caird et al. 2008 in the UK; Moglia et al. 2018 in Australia).

Of course, household issues of energy efficiency (in particular through heating and cooling, food storage and cooking, lighting and use of an increasing array of electronic equipment) intersect with national and global environmental concerns, including anthropogenic climate change. Australia is one of the world's largest emitters of greenhouse gases, especially on a per capita basis, and the country's established (as opposed to new) housing stock, which comprises the lion's share of the residential housing market at around 98 per cent (National Housing Supply Council 2013), is a significant contributor to these emissions (Harrington and Foster 2008, p. ix). In addition, a key feature of the Australian cultural context is a polarised and antagonistic climate change debate, intersecting with politicised traditional news media (Hackett et al. 2017; Chubb 2012, 2014; Bacon 2011). Indeed, insofar as the catalysts and barriers to improving energy efficiency are cultural—for home-making can be viewed as a resource mobilised behind different visions and values for of humanity and its future—this study provides a much-needed arena for public discussion of the big issues of our time.

Integrated into the social practice of home renovation is a diversity of media use as well as a plethora of everyday tactics deployed to help navigate the complexity of the renovation process, including purchasing decisions or dealings with building practitioners carrying out the job. Thus, while contemporary home renovations are mediatised in nature—that is, impacted by 'the everyday saturation of media and communication technologies' (Jansson 2017, p. 1)—what must be considered are the diverse dynamics and influences of media practice and engagement at the domestic and everyday level.

Although focused on media in home-making practices, the study we draw upon here follows a distinctive "media ecology" approach, plotting the relationships between different media types as well as media technologies and people (Podkalicka et al. 2016). Importantly, therefore, it locates media engagement as part of a broader suite of social practices (i.e. people's everyday habits and routines) 'focused directly or indirectly on media' (Couldry 2004, p. 117), corresponding to anthropological 'socio-centric' and 'non-media centric' approaches to media research. Tufte (2017, p. 263) explains this perspective as a 'holistic' and 'context-sensitive line of enquiry, which recognises that individual behaviour is deeply rooted in a complex set of socio-economic, cultural, political and spiritual contexts'. To examine media using this approach means situating their use 'in the context of communication ecologies and the broader socio-ecologies of everyday life' and in relation to 'social uses', which 'connects with behaviour and social change' (p. 263).

Insofar as the key aim of this collaborative research project was to provide empirical detail on the role of media in renovation practices, it was designed to locate the use of media within its everyday context of "mundane", urban domestic media consumption and production. Its geographical focus is the growing multi-cultural setting of Melbourne, Australia's second-largest city of some five million people in the nation's southeast, which is tipped to pip Sydney as the most populous city by the 2030s (Salt 2017), where the media environment of home renovations is multidimensional and plural, with digital media presumably "always on" or at least always accessible. While previous academic and industry research had identified the internet and social networks as significant for advice on renovating and energy efficiency (e.g. Office of Heritage and Environment 2014), exactly how the media—across traditional, digital and social media—figured in renovators' own research, decision-making and implementation of ideas was much less explored. To address this gap, the study adopted mixed-methods approaches including group discussions with Melbourne-based home renovators and an online survey of renovators across Australia (=156) (for further description of methods, see Hulse et al. 2015; Podkalicka et al. 2016).

The research revealed a wide range of media use at different stages of the renovation process—from relying on the internet and social networking sites such as Pinterest and Houzz for inspiration, through to using social media to post pictures during and after renovation. In the online survey, the internet was reported as the top choice (approximately two-thirds of respondents) for renovation ideas and inspiration, followed by print media (18 per cent). As described in earlier work (Hulse et al. 2015; Podkalicka et al. 2016), visual media came up as an important means for comparing designs, features, colours and so forth, while popular mainstream home improvement TV shows could provide design ideas, or at least become a point of conversation for building practices (often deemed "unrealistic") and consumer choices that featured on screen. Importantly, while the research confirmed the everyday saturation of digital media, with "Googling" a standard first-step renovation-research practice and with renovators switching to online versions of their favourite lifestyle magazines, we found diversity of media use and numerous inconsistencies. For example, some home renovators continued to be drawn to hard copies of retail home catalogues and a couple of participants admitted deactivating their Facebook accounts (Podkalicka et al. 2016, p. 15). As we explore further in this chapter, there remained a propensity to draw on personal networks to assist in accessing "trustworthy" information and building expertise, including for energy efficiency.

In this chapter, we are especially interested in exploring within these renovator discourses some of the relationships between local, national and/or international content and traditional and digital information sources, and their impacts on context-specific social practices and knowledge formation. Given its focus on media use in everyday life, the study encompassed the analysis of a wide range of media types, genres, cultural representations and communication forms, from the banal to the spectacular, the instrumental to the overtly political, the generic to the designated green: from instructional do-it-yourself (DIY) videos found on YouTube, to reality TV programmes such as Australian competition renovation show *The Block*, and from generalist interior design–focused lifestyle magazines such as *Habitat* to the specialist "green" magazines *Sanctuary* and *Renew*.

The empirical material selected for analysis here was derived from group discussions (=5) of approximately one and a half hours conducted with home renovators based in inner and mid-Melbourne. The participants represented homeowners, mostly aged between 30 and 50, with varied levels of renovation experience and living in different house types (including semi-detached units, apartments and detached houses). Many participants have accumulated their renovation experience through prior smaller-scale renovations and some applied their skills from cognate areas such as landscape designing or project management. While there was a diversity in terms of types and scales of renovations, popular areas for renovations were kitchens and bathrooms to 'modernise' and 'upgrade', which corresponds to the findings from our online survey that had them as the top answer with 35 per cent of respondents. Some were renovating to create a 'comfortable' family home long term, while some had a resale value in mind and were therefore planning to stay short term.

The focus group participants were recruited with the help from Melbourne local councils and the research team's professional networks, inviting people who, at the time of this research, had either implemented home improvements in the past three years or were considering renovating their home in the near future. The recruitment's focus was on the 'home renovator', rather than more narrowly on the 'green renovator'. The cohort was generally skewed towards well-educated middle-class demographic, with diverse attitudes towards sustainability/energy efficiency at home. These attitudes can be roughly mapped onto two groups of 'committed' greens (i.e. those prepared for additional personal expense for environmental gains) and 'material' greens (i.e. those environmentally aware but unwilling to pay for energy efficiency) (Newton and Meyer 2013). Despite the marked differences in participants' individual circumstances (and gendered experiences of renovation), the discussions revealed the common issue of cost as key to considering and implementing energy efficiency at home—this came through especially strongly when declaring the intention to make profit through the resell of the renovated property. Those with longer-term ownership plans tended to be more interested in the green transition, including in larger-scale changes such as the change of house layout to take advantage of the environmentally preferable north-facing orientation. Shared across the two different groups was also a perception of the need for more relevant, user-friendly information about energy-efficient products, methods and regulations and the high cost of professional advice, which the participating renovators reported as a barrier to the adoption of green products and techniques.

Local, Lived and Everyday

Importantly, the empirical material elicited from discussion groups clearly bears out the dynamics of media engagement, notably *digital* media engagement, as part of a broader suite of social practices anchored in local, lived, everyday experience—from face-to-face conversations with neighbours, colleagues and friends, to participation in community outreach environmental initiatives such as "speed-date an expert" schemes run by local governments or not-for-profit environmental organisations. A fundamental question asked of home renovators was what specific sources of information they drew upon for their renovation projects and this included when seeking to make improvements in home energy efficiency. The following response is representative of the breadth of information drawn upon by renovators:

> It was online searches. I definitely was *talking* to local manufacturers, the sellers of these products, so I was talking and *visiting* all of them. But all the other information about reliability and differences between products and

> then what's suitable for what I definitely wanted, yeah, it was really just searching around online and using online media, including discussion forums and other people's experience.

Intriguingly, here we find that locally defined social interactions, marked by the habits of "talking" and "visiting" (our emphasis, above), cannot be isolated from a range of digital media practices that are used to source or extend, check and validate knowledge. Despite differences in the individual design and combination of local versus digital practices employed to inform consumption decisions—for example, some renovators are generally more active media users than others—both digital media ***and*** face-to-face interactions affect local decisions and modes of consumption.

Here "talking" to people is viewed as an important practice carried out both in person and through digital and social media, by drawing on existing social contacts as well-generating new contacts through local, digital and social media communities. One key function of talk in this renovation context is to make general information personalised and specific (another function relates to trust; more on that in a moment), as the following example shows:

> Definitely, family and friends when they come over a lot. 'Look at this. Do you reckon this would be a really good one', particularly for us, we've been in that house six years and we still don't have a pergola, and I reckon he's built probably four for other people since we've been there, because we just can't decide on what we want to do.

Other participants sought online "talk" for design ideas using social media, with one renovator revealing that for them, for example, it was 'a lot of Facebook pages of the sorts of magazines that I like, like *Habitus [Living]*'.

Mediatised, Mediated and "Mundane"

While home renovations represent a case of "mediatisation" of everyday life, there is no clear-cut pattern of media use amongst renovators. Some participants, for example, insisted on the importance of "actually seeing" the renovations that other people had undertaken in their home (Podkalicka et al. 2016). They mentioned being proactive in creating these opportunities, walking around their local neighbourhood with the purpose of identifying a similar build or renovation project, and then asking its owners to share their renovation experiences:

A1: I have left notes in people's letterboxes saying I like [what they are doing]. People invite me into their house and say, 'Yeah come over and talk'.
Q: So you would actually leave a note saying, 'I'm renovating around the corner and …'.
A1: Yep with my email address, and they email me.
A2: That's a good idea.
A3: We actually get a lot of work that way: put a sign out the front of someone's house. We always get calls like that.

Another participant also spoke of visiting homes on the recommendation of a friend:

A: A friend who suggested a particular architect to me also said, 'Why don't you go and …'. He told me the location of some other houses that he knew were built by that architect, so I went and looked at one in Richmond and slipped a little note in their letterbox and asked if they would give me a call, which they did.
Q: And you were able to go and see the house?
A: She offered. I actually didn't have time to see it but she said they were similar to us: they knocked the back half of their house down, had a lot of work done to the front like restumping etc. I think at the time they moved out they had a two-week-old baby and a three-year-old or something.

"Popping into the neighbours' house" to have a look at what they have done is arguably a common practice. A noteworthy point here is how these "mundane" conversations are grounded in the lived experience of building a home or doing a renovation as a milestone project. They are thus mediated by context, and can include references to myriad other everyday conversations too. Also noteworthy is the way media content and its use are discussed and relayed amongst "insiders"—family, friends and social contacts—including through speech that contains the words of other people, whether quoted directly or paraphrased (Soep 2006). In this sense, the creative practice of home renovation situates the home renovator in ongoing dialogic relations with others. This could be viewed as a type of "crowded talk" (Soep 2006), where one's voice is never singular but rather an articulation of multiple voices, be they mediated or in-person.

In many of these narratives, traditional and digital media platforms occupy a significant, if itself "mundane", place in the ways in which people carry out their renovations (see above). However, against the backdrop of the presumed constant availability, accessibility and global reach of digital media, people continued to draw on their local knowledge base and social networks when it came to making renovation decisions. For example, some participants admitted to having drawn inspiration for interior design from a home renovation reality TV show in Australia, *The Block*. In this highly rating show, broadcast on the commercial free-to-air Nine Network, couples compete against each other to renovate houses and sell them for the highest price. *The Block* is one of many renovation programmes presented on Australian screens: from renovation competitions to general lifestyle programmes with segments devoted to home improvements and make-overs (e.g. *The Living Room* on the commercial free-to-air Network Ten). In various ways, these shows entertain and address the desires of homeowners as well as property investors, showcasing nifty designs, DIY and building skills and products for home-making or profit (e.g. Rosenberg 2008). Collectively, the multiplicity and the popularity of mediated renovations reflect and embed the cultural values that make Australia the 'Renovation Nation' (Allon 2008), with roughly 8.4 million (or 62 per cent) of Australia's 13.6 million homeowners involved in some type of renovations in 2016 (Roy Morgan Research 2017).

Reflecting on the relevance of *The Block*, some participants were adamant about not being directly swayed by the programme's ideas, instead emphasising the social, face-to-face context in which ideas permeate and spread:

> We had the friends who had done all the renovation, and then one of our neighbours, she's not an interior designer but her Mum was and she's really into it and she would constantly come around and go, 'Have you thought about this?' So that we'd get ideas from that and check them out.

Trust and Translation in Communication

The role of informal, word-of-mouth communication came through strongly in the study. Discussions with Melbourne-based home renovators confirmed the persistence, in the urban context, of informal relationships that shape communication and consumption, because of their local currency and trustworthiness (Podkalicka, et al. 2016; Podkalicka and Milne 2017). When asked directly if the local dimension was important, these participants pointed to the significance of trust:

A1: It is, because you might not trust somebody. For instance I would ring a [friend of mine] and ask who they recommend to do something and because they have a base of people that they know well, then I'd take on whoever they recommended, because I didn't have anybody here that could do that. Then to just pull up the Yellow Pages [business phone directory] and just choose somebody.
A2: Yes I'd rather hire someone under a recommendation rather than a nice fancy website.
A3: Yeah.
A1: Preferring to consult people first over information available through media.

Intriguingly, here discussion of the local is infused with discourse of the collective (of belonging to a community) without being in conflict with the individualism—'their ideas, their money' (Hulse et al. 2015, p 5)—that defines the dominant consumer culture of home renovation.

Another renovator, when expanding on their response about whether the local dimension matters to them, discussed "community" and knowledge of local council building regulations as key factors that impacted on decisions:

> I was about to say, I've done the same thing and I've sourced people from different areas, and the ones that were local, when they come they're just a bit more relaxed because it's their neighbourhood, it's their community, and I just found them a bit more understanding of what happened within that council. And I thought, well, if they're going to drive past my house every second or third day, they're going to take a bit more pride in what they're doing. So that was my slant to it.

What comes out strongly in this particular quote is the reference to the physical locality as a resource that engenders an intangible value of "pride", which in turn translates into expectations of quality work with material implications. The finding about the significance of local connections resonates with previous studies that have underscored the role local governments can play in promoting domestic energy efficiency because of being considered 'a trusted source of information' (unlike commercial energy companies) and having 'local knowledge of the needs of their residents and communities (which central government does not)' (Bale et al. 2013, p. 834).

As noted elsewhere, home renovators do a lot of their own research when contemplating renovation projects, seeking out information that is locally relevant, socially validated and includes face-to-face communication (Hulse et al. 2015; Podkalicka et al. 2016). They are navigating what is simultaneously abundant and fragmented information, with trust and pride viewed as paramount values and ingredients in the process of "translation" between renovation ideas, judgement and actual implementation. Indeed, these values and ingredients are often expressed as if self-evident. As one participant explained:

> We didn't quite know how to translate our ideas or our brief, so we did use our friends who said, 'Oh this builder', or I rang [someone] at the Victorian Master Builders [an association of practitioners intent on raising standards within the industry] and they said, 'Oh here's a builder who works locally', and we were sort of trying to make it a bit more sustainable, so we followed those networks.

Another participant elaborated on the importance of the social dimension, revealing just how much digital media practices are intertwined with those of traditional media (such as magazines), as well as how significant "speaking" to people is for making consumption decisions:

Q1: Would [you] just do a Google search?

A1: Yeah, do a Google search, or if you had a particular product in mind, you've read about it, and let's say, for example, in a *Renew* magazine, just looking at solar panels to install, so looking at different brands, honing that down to a number, and then actually contacting them directly and then speaking to them about it, and seeing what sort of engagement they have. A lot of the time, it's the people that actually respond, they're the ones you initially take up.

In the excerpts above, we can detect an unwillingness to accept "outside" expert advice, which is pitted against the perceived power of local or "inside" knowledge—and indeed the propensity to "engage" actively. In particular, these renovators note the role of personal networks and recommendations to access trusted practitioners and reliable information. Their narratives appear to reflect a perceived wider consciousness on issues of trust—what Bakir and Barlow have termed 'the generalised mood of modern times': that trust is waning (2008, p. 3). On the one hand, the study draws attention to people's sense of mistrust, or selective trust, in the media, an "institution" upon which they also rely for information, inspiration and ideas. On the other, the study emphasises how, in the hunt for experts on "home ground", the local takes on new significance in place- and identity-based responses to social and environmental problems of risk, sustainability and change (c.f. McKechnie 1996).

Study participants indicated this was similarly the case when using instructional DIY YouTube videos made, for instance, by people in the United States (Podkalicka and Milne 2017) as well as international reality TV programmes such as British *Grand Designs*. Further, while these visual media sources were viewed as entertaining and inspiring, they were also deemed as having a potentially limited purchase locally, given differences in product availability as well as practitioner skills (Podkalicka et al. 2018). Certainly, discussions with home renovators illustrate how media engagement remains a site of negotiation—where international, "territorially distant" content, whether used leisurely or purposefully, is "translated" to make sense within local circumstances.

The Interplay Between Local and Digital

As this study of home renovations and media in Australia demonstrates, the interplay between the local and the digital in environmental communication is tangled and complex. We find evidence of people translating ideas across local, national and global contexts—reconciling more abstract ideas received via both mediated and unmediated contexts with local experience, skills and knowledge. Indeed, the study reveals renovators responding to "human stories", which they can consider in the context of their own renovation ideas and circumstances (Hulse et al. 2015). The use of the internet and digital media, including online forum such as Whirlpool or visual social networking sites such as Pinterest, Instagram or design and home improvement app Houzz, was found to be a convenient and familiar means of seeking informal advice on the choice of reliable products or general renovation inspiration. Less commonly used were digital media to source trustworthy practitioners in local places—although there were participants who did admit to reaching out to their local Facebook groups for guidance on that, conjuring up the function of hyperlocal media as 'an alternative to more established local media outlets' (Rodgers 2018, p. 74). However, some renovators revealed that they had consulted local newspapers, such as *Yarra News* or *Melbourne* magazine, for content about local events and also to source building practitioners (Podkalicka et al. 2016).

Overall, the common reliance on the internet and digital media cannot be decoupled from what can be referred to as "unmediated" social interactions and communication. This point is well canvassed in media studies research, where digital media are seen as extending local and face-to-face interactions to larger virtual networks. This amplifies the effects of personal preferences. For example, as James Webster writes, 'social media often help people coalesce into like-minded communities leading them to media products and messages that resonate with group norms and, perhaps, reinforcing their predispositions' (2014, pp. 13–14).

That 'old' social and communicative practices persist in the digital environment is not a new finding, of course. Numerous commentators have identified the existence of the complex media ecologies as well as the processes of 'borrowing among media forms' called 'remediation' (Bolter 2007, p. 25). On the other hand, the 'mediatisation' research highlights complexities, 'inherent contradictions' and '*a dialectical relationship* between liberating forces and increasing socio-technological dependence' (Jansson 2017, p. 3; emphasis in original). This study thus adds to empirically grounded media research that grapples with the sheer complexity of configurations of media and communication practices and their integration into domestic practices and social life more broadly, here in the context of home renovations in Melbourne.

In this chapter, we have drawn out some of that complexity to explore the intersections between the local and the digital. One key finding relates to the experience of the local place, which, because of the nature of renovations, becomes a site of attuned visual exploration (via visual social media and by home visits) and social relations. While in many instances media assist home renovators in conducting research and actual renovations in the practical sense, the neighbourhood too offers systems of knowledge and advice anchored in existing and new social relationships. The common practices of "visiting neighbours" and "leaving notes in letterboxes" encapsulate traditional communal and predigital characteristics. That said, the "map" of the local environment can extend beyond the socio-spatial boundaries of the neighbourhood and into surrounding suburbs—as exemplified in the account by the home renovator interested in employing a particular architect—and also into *digital* spaces. The "local" is experienced first-hand when home renovators engage with practitioners, manufacturers or retailers face to face but it is also influenced by digital affordances in the form of locative media and digital maps. This locative media function is reflected in one participant's comment on the digital resources offered by local government (or shire councils):

> I find that various councils have got good skills in certain areas. For example, the City of Whitehorse [in the eastern suburbs of Melbourne] has got a fantastic plant database, and it's GPS-specific so you enter your address, and they'll produce something off the local map. Other councils are very generic; [a] lot of Bayside councils have got one generic plant list. So I think that some of them are good at certain things and others let them fall down. City of Port Phillip's got great materials on energy [just south of the CBD], just like the City of Yarra does [inner east and northern suburbs].

Another key finding is the combined use of the local and digital communication to build trust when making decisions about consumer products and desired building and design skills. In this case, the research provides evidence to support the argument that local places and digital spaces are interconnected—and both are resources used to navigate the abundance and fragmentation of available information. Lived experience is generated, valued and privileged as a useful source to guide people through the renovation process; indeed "local knowledge" forms a social construct infused with values and assumptions. This is further shaped by culture hence impacted by the everyday saturation of media and communication technologies, forms and content.

Implications for Environmental Communication

A key task for home renovators lies in navigating the domains of lived experience, expertise, resource and imagination, which they do by 'any media necessary' (Jenkins et al. 2016). Renovation processes span information-seeking and material transformation practices that are shaped by digital spaces and *simultaneously* demonstrative of an unrelenting relevance of the unmediated, informal, localised contexts embraced for environmental knowledge and practical support. These layers of media use and communication, encompassing different devices, face-to-face and mediated communication, attest to the 'deeper complexities and contradictions' in the everyday use of media as part of daily life (Tufte 2017, p. 262). This chapter sought to unpack such complexity and contradiction through an empirically grounded understanding of how digital media and communications technologies are implicated and experienced in the everyday material transformation of homes. It reveals the diversity of *situated* media use *and* everyday tactics deployed by home renovators to navigate renovation projects, in turn highlighting the nuanced interplay between the local and digital. In particular, it emphasises the persistence of 'old' social and communicative practices in the digital environment—notably, how "the local" takes on new significance in place- and identity-based responses to mediated, global, social-environmental concerns.

Although the study is focused on the media, however, it makes an active bid to complement (if not avoid) the media-centrism that defines much research on environment and communication. Rather, we advocate for building upon, or "grounding", models in media and environmental communication (markedly, the rich evidence provided by agenda-setting and framing models) within testimony of everyday, lived experience. This speaks to a broader proposition. In today's diversifying media landscape, how might we better understand human interaction, experience and agency in environmental communication—or what the cultural consumer "makes" or "does"—by embracing more integrated, culturally informed approaches to media research? By making media practices and the 'sub-processes' of mediatisation the foci of this chapter, in effect we are reconsidering the saliency of an "old" argument in cultural studies; as de Certeau (1984) wrote in his seminal text, *The practice of everyday life*, researchers need to build upon our knowledge of social phenomena by determining the use to which people put them. This, we propose, can only lead to a better understanding of the realities of media use and public perception of environmental concerns, not to mention people's capacity to act.

Acknowledgements The research underpinning this chapter was funded by the CRC for Low Carbon Living (LCL) Ltd. supported by the Cooperative Research Centres programme, an Australian government initiative. Thanks to collaborators: Kath Hulse, Esther Milne, Tomi Winfree and Gavin Melles; PhD candidates Shae Hunter, Aggeliki Aggeli and Sarah Fiess, as well as colleagues from a related CRC LCL project, especially Magnus Moglia and James McGregor.

References

Allon, F. (2008). *Renovation nation: Our obsession with home*. Sydney: University of New South Wales Press.

Allon, F. (2014). At Home in the Suburbs: Domesticity and Nation in Postwar Australia. *History Australia, 11*(1), 13–36. https://doi.org/10.1080/14490854.2013.11668498.

Andersson, E., & Öhman, J. (2017). Young people's conversations about environmental and sustainability issues in social media. *Environmental Education Research, 23*(4), 465–485.

Arniani, M., Badii, A., de Liddo, A., Georgi, S., Passani, A., Piccolo, L., & Teli, M. (2014). Collective awareness platform for sustainability and social innovation: An introduction. Report. http://booksprints-for-ict-research.eu/wp-content/uploads/2014/07/BS5-CAPS-FIN-003.pdf. Accessed 30 June 2018.

Bacon, W. (2011). A sceptical climate: Media coverage of climate change in Australia – Part 1: Climate change policy. Report. The Australian Centre for Independent Journalism and Global Environmental Journalism. https://www.uts.edu.au/sites/default/files/sceptical-climate-part1.pdf. Accessed 30 June 2018.

Bakir, V., & Barlow, D. (2008). The age of suspicion. In V. Bakir & D. M. Barlow (Eds.), *Communication in the age of suspicion: Trust and the media* (pp. 3–8). New York: Palgrave Macmillan.

Bale, C., McCullen, N., Foxon, T., Rucklidge, A., & Gale, W. (2013). Harnessing social networks for promoting adoption of energy technologies in the domestic sector. *Energy Policy, 63*, 833–844.

Bolter, J. (2007). Remediation and the language of new media. *Northern Lights, 5*, 25–37.

Brevini, B., & Murdock, G. (Eds.). (2017). *Carbon capitalism and communication: Confronting climate crisis*. London: Springer.

Caird, S., Roy, R., & Herring, H. (2008). Improving the energy performance of UK households: Results from surveys of consumer adoption and use of low- and zero-carbon technologies. *Energy Efficiency, 1*, 149–166.

Chubb, P. (2012). Really, fundamentally wrong: Media coverage of the business campaign against the Australian carbon tax. In E. Eide & R. Kunelius (Eds.), *Media meets climate: The global challenge for journalism* (pp. 179–295). Gothenburg: Nordicom.

Chubb, P. (2014). *Power failure: The inside story of climate politics under Rudd and Gillard*. Melbourne: Black Inc.

Connolly, J., & Prothero, A. (2008). Green consumption: Life-politics, risk and contradictions. *Journal of Consumer Culture, 8*(1), 117–145.

Couldry, N. (2004). Theorising media as practice. *Social Semiotics, 14*(2), 115–132.

de Certeau, M. (1984). *The practice of everyday life*. Berkeley: University of California Press.

Gibson, C., Waitt, G., Head, L., & Gill, N. (2011). Is it easy being green? On the dilemmas of material cultures of household sustainability. In R. Lane & A. W. Gorman-Murray (Eds.), *Material geographies of household sustainability* (pp. 19–33). Surrey: Ashgate Publishing.

Hackett, R., Forde, S., Gunster, S., & Foxwell-Norton, K. (2017). *Journalism and climate crisis: Public engagement, media alternatives*. Oxon: Routledge.

Haider, J. (2015). The shaping of environmental information in social media: Affordances and technologies of self-control. *Environmental Communication, 10*(4), 473–491.

Hansen, A. (2015). Promising directions for environmental communication research. *Environmental Communication, 9*(3), 384–391.

Hansen, A., & Cox, R. (Eds.). (2015). *The Routledge handbook of environment and communication*. Oxon: Routledge.

Harrington, L., & Foster, R. (2008). Energy use in the Australian residential sector 1986–2020. Report. Australian Department of the Environment, Water, Heritage and the Arts. http://www.energyrating.gov.au/sites/new.energyrating/files/documents/2008-energy-use-aust-res-sector-full%5B1%5D.pdf. Accessed 30 June 2018.

Hepp, A. (2013). The communicative figurations of mediatised worlds: Mediatisation research in times of the 'mediation of everything'. *European Journal of Communication, 28*(6), 615–629.

Housing Industry Association. (2017, July 6). *Home renovation market gets a makeover*. Press release. https://hia.com.au/-/media/HIA-Website/Files/Media-Centre/Media-Releases/2017/national/Home-Renovation-Market-Gets-a-Makeover.ashx. Accessed 30 June 2018.

Hulse, K., Podkalicka, A., Milne, E., Winfree, T., & Melles, G. (2015). 'I'd just Google it': Media and home renovation practices in Australia, Melbourne. Report. CRC for Low Carbon Living. http://www.lowcarbonlivingcrc.com.au/sites/all/files/publications_file_attachments/rp3021_project_report2016.pdf. Accessed 30 June 2018.

Jansson, A. (2017). *Mediatisation and mobile lives: A critical approach*. London: Routledge.

Jenkins, H., Shresthova, S., Gamber-Thompson, L., Kligler-Vilenchik, N., & Zimmerman, A. (Eds.). (2016). *By any media necessary*. New York: NYU Press.

Judson, E. P., & Maller, C. (2014). Housing renovations and energy efficiency: Insights from homeowners practices. *Building Research and Information, 42*(4), 501–511.

Koteyko, N., Nerlich, B., & Hellsten, I. (2015). Climate change communication and the internet: Challenges and opportunities for research. *Environ-*

mental Communication, 9(2), 149–152. https://doi.org/10.1080/17524032.2015.1029297.

Lester, L., & Hutchins, B. (2009). Power games: Environmental protest, news media and the internet. *Media, Culture & Society, 31*(4), 579–595. https://doi.org/10.1177/0163443709335201.

Lester, L., & Hutchins, B. (2012). The power of the unseen: Environmental conflict, the media and invisibility. *Media, Culture & Society, 34*(7), 847–863. https://doi.org/10.1177/0163443712452772.

Maibach, E., Nisbet, M., Baldwin, P., Akerlof, K., & Diao, G. (2010, June). Reframing climate change as a public health issue: An exploratory study of public reactions. *BMC Public Health*. https://doi.org/10.1186/1471-2458-10-299.

Maller, C., Horne, R., & Dalton, T. (2012). Green renovations: Intersections of daily routines, housing aspirations and narratives of environmental sustainability. *Housing, Theory and Society, 29*(3), 255–275.

McKechnie, R. (1996). Insiders and outsiders: Identifying experts on home ground. In A. Irwin & B. Wynne (Eds.), *Misunderstanding science? The public reconstruction of science and technology* (pp. 126–151). Cambridge: Cambridge University Press.

McKeith, S. (2016, February 4). Buyers swap 'traditional Aussie dream' for high density apartments. *Huffington Post*. https://www.huffingtonpost.com.au/2016/02/04/high-density-apartments_n_9145014.html. Accessed 30 June 2018.

Moglia, M., Podkalicka, A., & McGregor, J. (2018). An agent-based model of residential energy efficiency adoption. *Journal of Artificial Societies and Social Simulation, 21*(3), 1–3.

Morris, M. (1990). Banality in cultural studies. In P. Mellencamp (Ed.), *Logics of television: Essays in cultural criticism* (pp. 14–43). Bloomington: Indiana University Press.

Napp, T., Shah, N., Fisk, D., & Clark, D. (2012, June 8). What's energy efficiency and how much can it help cut emissions? *The Guardian*. https://www.theguardian.com/environment/2012/jun/08/energy-efficiency-carbon-savings. Accessed 30 June 2018.

National Housing Supply Council. (2013). 2013 State of supply report: Changes in how we live. Australian Department of Treasury. https://static.treasury.gov.au/uploads/sites/1/2017/06/NHSC-State-of-Supply-Report-Consolidated.pdf. Accessed 30 June 2018.

New South Wales Office of Environment and Heritage. (2014). Sustainable Households: Survey of homeowners. Report. http://www.environment.nsw.gov.au/resources/actionmatters/IR-sustainable-households-survey-2014.pdf. Accessed 30 June 2018.

Newton, P., & Meyer, D. (2013). Exploring the attitudes-action gap in household resource consumption: Does 'environmental lifestyle' segmentation align with consumer behaviour? *Sustainability, 5*(3), 1211–1233. https://doi.org/10.3390/su5031211.

Nisbet, M. (2009). Communicating climate change: Why frames matter for public engagement. *Environment: Science and Policy for Sustainable Development, 51*(2), 12–23. https://doi.org/10.3200/ENVT.51.2.12-23.

Nisbet, M., & Kotcher, J. (2009). A two-step flow of influence? Opinion-leader campaigns on climate change. *Science Communication, 30*(3), 328–3054.

Podkalicka, A. (2018). Actor, intermediary, and context: Media in home renovation and consumption practice. *Communication Research and Practice, 4*(2), 210–225. https://doi.org/10.1080/22041451.2018.1507326.

Podkalicka, A., & Milne, E. (2017). Diverse media practices and economies of Australian home renovators: Budgeting, self-education and documentation. *Continuum: Journal of Media & Cultural Studies, 31*(5), 694–705.

Podkalicka, A., & Rennie, E. (2018). *Using media for social innovation*. Bristol: Intellect.

Podkalicka, A., Milne, E., Hulse, K., Winfree, T., & Melles, G. (2016). Hashtag sustainability? Home renovators' media world. Report. CRC for Low Carbon Living. http://builtbetter.org/node/3196. Accessed 30 June 2018.

Podkalicka, A., Milne, E., & Kennedy, J. (2018). *Grand Designs: Consumer markets and home-making*. London: Palgrave Macmillan.

Rodgers, S. (2018). Digitising localism: Anticipating, assembling and animating a 'space' for UK hyperlocal media production. *International Journal of Cultural Studies, 21*(1), 73–89.

Rosenberg, B. C. (2008). Property and home-makeover television: Risk, thrift and taste. *Continuum: Journal of Media & Cultural Studies, 22*(4), 505–513.

Roy Morgan Research. (2017). Renovation nation: Home improvement in Australia. Press Release 7102. http://www.roymorgan.com/findings/7102-renovation-nation-home-improvement-in-australia-201701090848. Accessed 30 June 2018.

Salt, B. (2017, May 25). Melbourne set to overtake Sydney. *The Australian*. https://www.theaustralian.com.au/business/bettercities/melbourne-set-to-become-nations-most-populous-city-by-2030s/newsstory/59ab02029829655b7be9e894a0133cbc. Accessed 30 June 2018.

Sheppard, J., Gray, M., & Phillips, B. (2017). ANUPoll – Attitudes to housing affordability: Pressures, problems and solutions. Report 24. Australian National University College of Arts and Social Sciences. http://politicsir.cass.anu.edu.au/sites/default/files/docs/ANUpoll-24-Housing-2017_0.pdf. Accessed 30 June 2018.

Soep, E. (2006). Beyond literacy and voice in youth media production. *McGill Journal of Education, 41*, 197–213.

Strengers, Y., & Nicholls, L. (2017). Convenience and energy consumption in the smart home of the future: Industry visions from Australia and beyond. *Energy Research and Social Science, 32*, 86–93.

Sustainability Victoria. (2014). Victorian households energy report 28. http://www.sustainability.vic.gov.au/services-and-advice/households/energy-efficiency/toolbox/reports. Accessed 30 June 2018.

Takahashi, B., Edwards, G., Roberts, J. T., & Duan, R. (2015). Exploring the use of online platforms for climate change policy and public engagement by NGOs in Latin America. *Environmental Communication, 9*(2), 228–247.

Tufte, T. (2017). *Communication and social change: A citizen perspective*. Cambridge: Polity.

van der Schoor, T., & Scholtens, B. (2015). Power to the people: Local community initiatives and the transition to sustainable energy. *Renewable and Sustainable Energy Reviews, 43*, 666–675.

Webster, J. (2014). *The marketplace of attention: How audiences take shape in a digital age*. Cambridge: MIT Press.

Index[1]

[1] Note: Page numbers followed by 'n' refer to notes.

J. Díaz-Pont et al. (eds.), *The Local and the Digital in Environmental Communication*, Global Transformations in Media and Communication Research - A Palgrave and IAMCR Series, https://doi.org/10.1007/978-3-030-37330-6

The manufacturer's authorised representative in the EU is Springer Nature Customer Service Centre GmbH, Europaplatz 3, 69115 Heidelberg, Germany. If you have any concerns regarding our products, please contact ProductSafety@springernature.com

Printed and bound by CPI Group (UK) Ltd, Croydon, CR0 4YY
06/10/2026
02228468-0004